DANA DASH:
FIRST GIRL ON THE MOON

S.S. HUDSON

ELEMENTIAD

This is a work of fiction. Names, characters, businesses, places, events, locales, and incidents are either the products of the author's imagination or used in a fictitious manner. Any resemblance to actual persons, living or dead, or actual events is purely coincidental.

ISBN 978-1-951421-00-7 (Paperback Edition)

Library of Congress Control Number 2019952349

Editing by Erin C. Tolman

Cover by Pat Redding Scanlon and Pinguino Kolb

Internal illustrations by Kathleen Wallace

Book design by Amy Kohtz and Idria Barone Knecht

Printed and bound in USA

First Printing October 2019

Published by Elementiad, Inc.

P.O. Box 1058

Portsmouth, NH 03802

Visit www.elementiad.com

For Dana, who wanted a book with her name in it,
and Bosco, who seemed very excited about this project
(and everything else)
and with all due apologies to H.G. Wells.

"Any sufficiently advanced technology is indistinguishable from magic."

Arthur C. Clarke ("Prophet of the Space Age") 1917 - 2008

(Author and Futurist. Possibly the first to imagine placing communications satellites at geostationary orbital distance, sometimes called Clarke Orbit or The Clarke Belt.)

"A ship in port is safe, but that's not what ships are built for."

Grace Hopper ("Grandma COBOL") 1906 - 1992

(Computer Scientist and U.S. Navy rear admiral. Developed one of the first high level computer languages and coined the computer science term "compiler."[0])

---

[0] A *compiler* is a special computer program that turns a high level, human readable (for some definition of human) computer language into machine code that a computer can execute.

CHAPTER 0 — DISASTER PLAN

It was not ten-year-old Dana Dash's intent to become one of the most important human beings in all of history — not right at that moment, anyway. She was thinking about only one thing: how she was going to explain this to her parents.

Dana was sitting in the principal's office. She was soaking wet from the sprinklers that had gone off in the science lab, and a puddle was forming on the dull gray carpet beneath her uncomfortable plastic chair. Her brown wavy hair was caked with plaster dust, and there were white streaks of it running down her face, shirt, pants, and shoes. She was waiting for Principal Peters to return from the lab, where the bomb squad had been called in from New York City to verify that the explosion had only been an air pressure demonstration gone awry and not something more nefarious. Her parents had been called, and her dad would be there any minute.

It wasn't fair. It wasn't her fault that her science fair project had exploded. It wasn't her fault that the school had been evacuated. It wasn't even her fault that the entire Dash-on-Hudson Volunteer Fire Department had been mobilized to the school or that the mayor had held a press conference on the playground of Hillhurst Elementary to dispel rumors of a terrorist attack while kindergarteners huddled in the background crying and waiting for their parents. It really wasn't her fault. Not this time.

The smug smirk on Matt McCaws's face — Dana's jaw clenched thinking about it. He'd sabotaged her experiment. She could've been killed. She was probably going to get expelled. Nobody believed her when she protested her innocence and pointed at the mayor's son. He definitely did it on purpose to ruin her experiment, although she doubted he knew how badly it would end: with a deafening bang as the steel drum exploded and shot upwards, blowing a hole in the ceiling, filling the room with steam, and knocking one of the fire safety sprinkler's caps loose. Everybody heard the bang. Somebody saw the steam and the dust and the sprinkler and pulled the fire alarm, and the whole school ran screaming and crying and pushing and shoving out the exits, in no way resembling the calm, orderly safety drills they practiced.

Dana had tried to get a teacher's attention to say it was an accident, but there was already too much chaos. And nobody listens to a ten-year-old girl.

But Principal Peters, who had grabbed his Incident Commander hard hat from his office wall where it hung at all times (this wasn't the first time Dana had sat in his office), listened to Matt when he meekly tugged on his sleeve and with innocent eyes pointed to Dana. Of course the principal listened to Matt; Matt was his nephew and the mayor's son. Principal Peters had grabbed Dana by the arm and frog-marched her to his office, bellowing about his *duty*, by *law*, to enforce school safety. He wouldn't even let her best friend Noah Knight wait with her, unless Noah wanted to be "guilty by association." The principal had hollered that phrase as he slammed the door to his office and stomped away to bark orders at the firefighters and police officers and teachers and anybody else he could.

So Dana waited — cold, wet, covered in goopy white muck, and fuming. The minutes dragged on.

There was a soft tap on the door. "Knock knock," a voice said.

Dana wasn't sure what to do. "Uh, who's there?" said Dana, thinking this was the most inappropriate time for a knock-knock joke she could imagine.

In response, the door opened, and her science teacher Ms. Astrulabi entered and quietly closed the door.

"Oh Dana," she said, shaking her head, her soft green headscarf swaying with the motion. "I'm not sure there's anything I can do to help you this time."

"But Ms. Astrulabi! It wasn't my fault! My experiment was safe, I showed you last week with the soda can!" Dana protested, feeling her face flush with indignation. "It was Matt. You know he has it in for me. He sabotaged the barrel — he put the cap back on while I wasn't looking!"

"I believe you, Dana. But alas, there's quite a scene outside. Why did you decide to use a larger barrel? The soda can demonstrated the power of air pressure quite well. And," she added, "was quite safe."

"I know, Ms. Astrulabi. But I wanted to be more impressive for the judges. A dinky little can crumpling isn't the same as a steel barrel imploding. I… I just wanted a good project for the science fair. I wanted to win!"

Dana's eyes began to fill with tears. Ms. Astrulabi was right. This wasn't like the time Dana let the mice out of the cage. Or when she "borrowed" the school's telescope outside to get a better look at the super moon. This time, although she had taken all the proper safety precautions,[1] she hadn't protected herself from the human element: sabotage. She hadn't run the full experiment — she was only there during lunchtime to test how long it took to boil the water in the larger barrel. How could she know that slimy Matt McCaws would want to win the science fair competition so badly he would follow her into the lab? He must have heard her say *"Make sure cap is removed — check!"* while she read her safety list. He might not have even known it would explode — he just wanted her to fail. And he was going to get his wish.

Dana thought about the scared kids and frantic parents and the news crews and the bomb-sniffing dogs and the emergency responders from at least five city, state, and federal agencies outside. This was very, very bad.

Down the hallway, Dana could hear voices and the heavy steps of Principal Peters. He thrust open the office door and sauntered behind his desk, still wearing his Incident Commander hard hat. It was too small for his big head and rested on top of his toupee like a cherry on an angry orange sundae.

Donald Dash, Dana's father, walked in behind the principal.

"Bug! Are you ok?" he said, and held open his arms.

---

[1] When conducting potentially dangerous experiments, first put on safety goggles and protective gear, and make sure a fire extinguisher is handy. Always. Safety first!

Dana jumped into his big embrace. She was really going to cry now. "Errrrmuuhhkaaay," she mumbled into his shoulder.

Principal Peters cleared his throat with a *Hrrrrumpff*, and Mr. Dash let go of Dana, smoothing her messy hair as he took a seat. Dana wiped her nose on her sleeve and braced herself. As bad as her day had been so far, she was pretty sure it was about to get worse.

"Hrrrrumpff." Principal Peters cleared his throat again. "Mr. Dash, I have made my official report to the superintendent. This is a very serious matter. And there will be serious consequences."

"What exactly happened?" Mr. Dash asked, turning to Dana.

"I was only..." Dana started, but Principal Peters cut her off.

"What happened was the consequences of little girls conducting dangerous science experiments!" Principal Peter's jowls wobbled as he tried to control himself.

"Dad, it wasn't..." Dana tried again to explain her side. Principal Peters was determined not to let her.

"Your daughter is a menace! Look at the... the *havoc*! Your daughter *blew up a school*!" He was gripping the edge of his desk so hard his knuckles were white.

"Dana, is this true?" her father asked, gently. Her father never raised his voice, but when he spoke like this, he was very, very disappointed. Dana's heart sank. He thought it was her fault.

"No, I... Dad, it wasn't me! Matt McCaws..."

"WAS IT OR WAS IT NOT YOUR BARREL THAT EXPLODED IN THE SCIENCE LAB?!?" Principal Peters bellowed.

"Well yes but..." Principal Peters interrupted before Dana could explain further.

"AND DID YOU OR DID YOU NOT HAVE PERMISSION TO DO THIS EXPERIMENT?!?"

"Well no..." Dana sunk into her chair. She couldn't argue that point. She hadn't asked.

"FIRE DEPARTMENTS FROM THREE NEIGHBORING VILLAGES!" Principal Peters' face was red and his eyes were starting to bulge, like he was conducting an air pressure experiment on his face.

"NEW YORK STATE EMERGENCY MANAGEMENT!" Collections of spittle were forming in the corners of his mouth.

"DEPARTMENT OF HOMELAND SECURITY!" His hairpiece was slipping from the effort of keeping the Incident Commander hard hat in place.

"THE COAST GUARD!" His hairpiece gave up and slipped loose down his forehead. The hard hat tumbled to the desk, bounced once with a plastic-y ***THWACK***, and landed at Dana's feet. She picked it up, and would have handed it back to the principal, except he hardly seemed to notice. He pushed his hairpiece mostly back in place and continued.

"I cannot overemphasize the embarrassment this has brought down on my head! And of course, the disruption to the school!"

"Was anybody hurt?" Mr. Dash asked.

"That," Principal Peters growled like a bulldog, "is not the point! The news is here! The mayor! And if it weren't for my nephew having the... the *courage* to say something because he saw something, we would have thought it was terrorists!"

He pointed to a sign on the wall that was repeated at every school assembly and safety drill: *If you see something, say something.* She hadn't actually seen Matt place the cap back on the barrel, but she had seen him running out the science lab's back door, just as the barrel started making loud ***PINGS.*** The experiment was supposed to show the power of air pressure by lowering the pressure inside the barrel and causing a very loud, but mostly safe, implosion. Dana's plan was to do this very dramatically for the judges, with a velvet curtain staged behind her, and pictures of moon landers to show an example where air pressure was no joke. They would have pinned the first place ribbon on her right there.

She was definitely not going to win now.

Dana realized she had tuned out the frothing bellows of her principal as he described to Mr. Dash the events of the day. "...activated the incident command system... threat assessment... bomb squad... terrorist watch list..."

"Wait, what?" asked Dana.

"... tell the Department of Homeland Security to put you on the terrorist watch list, little girl!"

"Dad, can he really — "

Mr. Dash shook his head, as if to say, *no, of course he can't, but also, this isn't a good time to question this man's authority.*

"... your permanent record!"

Mr. Dash leaned forward in his chair. Dana could see he was at the end of the amount of time he liked to spend talking to people.

"What exactly is a permanent record for a ten-year-old, Principal Peters?" he asked.

"It is — I'm going to — she will... be banned from science!" Principal Peters sat back triumphantly in his chair and made a noise, probably the chair cushion sighing from the large man shifting his weight, but it sounded like a fart.

Hearing this — the punishment, not the fart noise — Dana wanted to both laugh and cry. *You can't ban me from science!* Ridiculous. But oh no... what if he said that she...

"...will be banned from entering the science fair..."

Dana felt sick.

"...and from performing science lab experiments for the rest of the year..."

Ok, she had seen that one coming.

"...and from participating in all extracurricular activities..."

No astronomy club either? They were doing a lunar observation project right now. What did he think she was going to do, blow up the moon? Dana looked at Ms. Astrulabi, the club's advisor. She gave Dana a sympathetic look.

"...and I will be speaking to Principal Snodgrass at Hillhurst Middle School so that these bans continue. It's no punishment if a child blows up the school two weeks before school ends and then goes home for the summer — scot-free to a new school next year — No! There will be *discipline!* And *order!* I have to protect my reputation... uh... the school. Protect the school from this... this... *willful* and *reckless* girl! She will learn that young ladies are supposed to follow rules! I'm sure Principal Snodgrass will agree with me that troublemaking girls have no place in a science laboratory!"

Dana felt numb. She couldn't feel her cold, wet jeans clinging to her legs or the ceiling plaster dust caked to her face. Just numb. No science labs. No science fairs. No science clubs of any kind. For the rest of her schooling, if Principal Peters had his way.

No science.

At that, Dana's father stood up. He looked like he was gritting his teeth.

"Excuse us, Principal Peters. We're going home now."

"Excuse *me*, Mr. Dash, I wasn't finished — " He cleared his throat again and resumed his rant, which Dana realized he certainly had practiced many times in his head. "And furthermore, Dana will be suspended — "

"Excuse *me*, Principal Peters. No. My daughter will be in school tomorrow," said Mr. Dash.

That was not the response Principal Peters was expecting. "Iffhh whaaa?"

"You base this suspension on the word of another student. A student to whom you are related. You want to suspend my daughter, without even the pretense of listening to her version of events."

"Yes, well, yes..." Principal Peters didn't seem to know where Mr. Dash was going with this, but he was pretty sure he didn't like it.

"My daughter may be many things — things that you, or others, might not want her to be — but she is not a liar. She admits using the science lab without authorization. She admits failing to get permission for the experiment. It is logical that she receive punishments correlated to those actions.

"But the gross negligence of your emergency response plan and the subsequent chaos is your fault, not Dana's. And her role in, not the possession of the barrel, but the *explosion* of the barrel, is based on the word of a boy who is a known bully. If you deny her her rightful access to education because of an unreliable witness, and scapegoat her to ease your own embarrassment, then I'm certain the Channel Four News team outside would relish an interview with Dana as the star witness to a school explosion. And how you stuck her in your office rather than have her checked by paramedics. And how the mayor's son is the real terrorist. Which should play well with the voters in your brother-in-law's primary election next month, don't you think?"

Dana's jaw dropped about halfway through her father's speech. She had never heard him say so many words so forcefully before. He was not a small man, but he had never loomed this large. He seemed to have been made taller by the righteousness of his indignation. His sudden defiance was another thing the principal was not prepared for today.

Principal Peters sat in his farty chair, flap-jawed and bulge-eyed, like somebody had slapped him in the face with a fish. Or maybe if one fish slapped another fish. *Yeah, that's what one fish slapped with another fish would look like,* she thought: completely, utterly astonished and confused by what had just happened.

Ms. Astrulabi, who stood silently against the wall, was biting her lip. Was she trying not to laugh?

Mr. Dash put out his hand. Dana stood up and took it, still holding Principal Peters' Incident Commander hard hat. She looked at the principal and weighed her options. Throwing it at his fat head, much as she'd like to, probably wasn't a good idea. Her father had just argued her out of a suspension. Better not to give them another reason.

On instinct, she put it on her own head.

Dana and her father walked out of the principal's office. She glanced behind her and saw Ms. Astrulabi failing at her attempts to stifle her laughter. And she saw Principal Peter, still dumbstruck, fish-face-fish-slapped, or perhaps like he had swallowed one of those fish whole.

Mr. Dash let her keep the hat on until they had left the building, then took it gently from her head and put it in her hands. Outside, people were still milling around the school. With no actual emergency, it had turned into an opportunity to chat with neighbors and schedule playdates while the kids ran around the playground like the whole thing was one extended recess.

They were about to get in the car when Ms. Astrulabi rushed across the parking lot.

"Dana!" She was waving something in her hand. A blue piece of paper, with a crest at the top.

"Ms. Astrulabi, I'm so sorry I can't be in the Astronomy Club anymore..."

"Never mind that." She paused as she reached them, a little out of breath, headscarf slightly askew. "I have something for you."

She thrust the paper into Dana's hands. Dana, who hadn't known what to do with Principal Peter's precious hat, handed it to Ms. Astrulabi in exchange.

"What is it?" she asked, looking at the paper. It looked like a flyer of some kind.

"It's a contest. Another science fair. Over the summer. At a different school." Ms. Astrulabi looked more excited than Dana thought she should be, considering that Dana could no longer be president of the Astronomy Club.

"A different school?" Mr. Dash asked.

"A science and technology school for girls... well, girls and boys. A private school. No connection to this principal or the mayor or anybody else who might have, shall we say, undue prejudices. Dana can apply through this competition."

"Oh," Mr. Dash said, shaking his head. "The Cavor School. We looked into it when Dana was younger; she went to a preschool there for a while. But private school — it's not something we can afford."

"But," said Ms. Astrulabi, "the competition prize is a scholarship." Her dark eyes were practically dancing.

A scholarship...

"Considering the state of affairs here," said Mr. Dash, eyeing Dana's current school, "It's worth a look."

They thanked her, and Dana opened her car door, eager to get away from this school and this awful day. She startled a little lizard sunning itself on the car roof. It made a valiant leap to the pavement and scuttled swiftly under another car and out of sight.

As they drove away, Dana leaned her head against the window, gazing up at the warm spring sunlight streaming through the tree branches. The terribleness of the events played over and over in her mind: the explosion, the chaos, her punishment. But as she looked at the flyer her favorite teacher had given her, the gears in her head began to turn. Dana knew that most kids her age would have still been reeling from a disaster like this, probably wanting to take a break for ice cream or candy while they calmed down. But not Dana. Dana simply cleared her mind, letting the stress and negative feelings drain away, and began working on a new plan.

She certainly wouldn't say no to the candy, though.

From the notebook of Moses Moreau

CHAPTER 1 — RACING TO WIN

Dana whooped "Kiai-yah!" and jumped over a large pothole rather than lose precious seconds going around it. In the air, her right leg flashed out in a flying *mae geri* karate kick to the head of her imaginary opponent. She landed back on her wheels.

It was Friday afternoon, one day after the explosion at the school. She and Noah were skateboarding, speeding down the steepest part of Big Oak Drive, in the village of Dash-on-Hudson. Dana's hair streamed out from under her helmet in a wavy brown tail that did not entirely unresemble a flying mop. Noah followed close at her heels, a look of steely determination in his eyes, his short black hair windblown in an aerodynamic mohawk. He was gaining on her, but Dana intended to win.

As they reached a stop sign, an expensive looking SUV pulled into the intersection. Dana launched herself over the hood of the luxury vehicle as it screeched to a stop, one hand on the hood as she vaulted over it like a streetwise gazelle. Her electric skateboard whistled under the engine of the car, narrowly missing the front tires, and emerged on the other side to meet her feet, as if it had been in radio contact with her shoes. The car gave a long angry blerp of its horn behind her but the sound barely registered as Dana continued on her mission.

Noah had chosen a path behind the car and onto the sidewalk. It was a safer calculation, but one that cost him valuable seconds. He pushed hard to make up the time, even closing the gap a bit on the last straightaway, but it was not enough.

"I win!" Dana exclaimed, arms thrust above her head in a victory V as they coasted into the Cavor School grounds at the bottom of the hill.

"Queen of Skate Mountain! Dash by name, I dash by nature![2]"

"Don't be..." said Noah as he caught his breath, "...too comfortable with your crown. I might beat you next time. If you hadn't almost gotten yourself killed with one of your insane stunts..."

"Never!" interrupted Dana, continuing her excessive celebration. "I shall forever remain the undisputed champion!" She spun a theatrical 360-degree turn on her skateboard, ending her triumphant twirl face to face with the imposing old house at the center of the Cavor School campus.

Myrtlegrove Manor's aging architecture was not unusual for the village of Dash-on-Hudson, which was part of a constellation of colonial villages[3] that dotted the Hudson River north of New York City. Dana's father had told her stories about Magnus Erisey Dash, the infamous founder of her hometown. He was a river pilot, an explorer, a Revolutionary War hero, and her many-many-many-greats grandfather. He was also, quite probably, a smuggler and a pirate, though this last point had been hotly debated.

Dana Dash, possible descendant of a pirate, gazed up at Myrtlegrove Manor.

"You're sure, when you called the school, they said to come to the old manor house, and not to one of the main school buildings over there?" asked Noah, pointing to a group of large, very much more modern and official-school-administrator-looking structures.

"That's what they said," Dana replied, looking from the cluster of buildings that seemed to sparkle in the late spring sun, glints of light winking off the windowed, white exteriors, and then to the aged stone mansion, first floor nearly overtaken by renegade ivy, thick curtains drawn across nearly every window. "They already mailed out the contest forms,

---

[2] Dana had stolen her catch phrase from a character with the last name Swift in a series of old books she found on her Dad's bookshelves and enjoyed reading very much.

[3] Irvington was named for the writer, Washington Irving (1783-1859); Tarrytown and Sleepy Hollow were locations for his stories of hauntings, curses, and hoaxes: Rip Van Winkle, the Headless Horseman, and Dietrich Knickerbocker.

but said I could still pick them up directly from the Head of School's office."

"You would think that a boarding school specializing in science and technology would have the forms online," said Noah.

"Yeah, really," said Dana. "I guess it's an old-fashioned boarding school first and a science and technology school second. It's something like a hundred and fifty years old."

"It definitely looks like it," said Noah, staring at the strange old house. "Looks they would take their traditions seriously."

Dana also found herself gazing at the architecture. She was particularly fascinated by a small green domed structure on the east side of the roof. It was like a small gazebo that only one person could stand in, or maybe two if they were very friendly. *Must be solid copper, with a patina that green*, Dana thought. A short walkway on the roof led from a small attic door to the odd copper cupola. Dana's mother had pointed out such walkways on the roofs of various large houses in the area. "Widow's walk, it's called," and she had described how the wives of seafaring men would go to the roof and gaze out at the sea, hoping each day to see their husband's ship on the horizon. But Dana had never seen a dome like this one. *Maybe*, she thought, *this particular wife got tired of waiting in the rain.*

"When is the deadline again?" asked Noah, interrupting her thoughts.

"Monday," said Dana.

"Monday, as in Monday the seventeenth?" Noah asked. Dana nodded. "Yikes, Dana. It's already Friday the fourteenth. That doesn't leave a lot of time to come up with a project idea and write a proposal," said Noah.

"Yeah. But I've already got a few ideas." She furrowed her brow at the thought of her project notebook, still blank. "Anyway, thanks for coming with me. It would have been a little scary to come here all by myself."

"Ha," said Noah. "You? Scared? I've never seen that. Besides, I want to go here as much as you do. And no, not just so I don't have to face middle school next year without you."

Dana laughed at this as she removed her skating helmet and put it in her backpack. She pulled out her project notebook and a pencil. It was a little ritual for her. Every new project needed a fresh, quad-ruled lab book to begin it right. And, in her mind, this project started right here, right now, in the shadow of this dusty old house.

On the front of the lab book she wrote *Cavor School STEM Contest.*

The campus road deposited them at the manor house, but it wasn't immediately clear where the entrance was. Several large stairways led from the road up to sprawling porches, lined with heavy stone columns and stained glass windows, but no doors. They walked around the southern side of the building, down a wide brick pathway with a small garden off to one side. Newly planted, with seedling sprouts emerging from mounds of dark soil, dotted with popsicle sticks with the names of the plant written in scrawled pencil, it was a quaint contrast to the imposing grandeur of the mansion. It even had a shabby scarecrow in a floppy sun hat staked at the head of the garden, looking more like a cheerfully straw-stuffed conductor of a vegetable orchestra than an ominous presence capable of striking terror into passing birds. Dana wondered who at the school would tend a garden like this and hoped it wasn't the students. She opened her mouth to say so to Noah, but could see he was eyeing the garden with interest and likely thinking the exact opposite.

They rounded the corner of the manor house. A circular drive wound around to meet the wide, curved front steps that intersected the porch. The only doors they could see were the pair before them — heavy, thickly carved dark wood, with large brass rings dangling from the front of each like earrings. They walked up the steps.

"Uh, should we..." Noah said, reaching his hand up to grab one of the brass rings. The doors dwarfed the two of them, and the sidelight window next to the doors that would have given them a glimpse into the entryway were mirrored — perhaps not even windows at all.

"Hmmm," said Dana, taking a step back towards the stairs and glancing around. "Ah! Look!" She pointed at a small, luminescent white circle just above the ramp, below what looked like a keycard reader. "Doorbell," she said. The doorbell was nestled in a small frame, with a carved crescent on either side.

She pushed it, and felt small ridges beneath her fingertip. Looking closer, the white doorbell seemed to have, unless Dana was very much mistaken, the same markings as a full moon on a clear night. Dana could be mistaken about many things — a good scientist has to be willing to make mistakes — but she was very rarely wrong about the moon. This house had a moon doorbell. *Cool,* she thought, and wondered if anybody inside would know where she could get one for her house. Was there anybody inside?

"Hey, put your ear up to the door and listen if the doorbell is even working," she said to Noah.

Noah pressed his ear up to the wood. "Ok, try again."

She pushed against the moon again, this time with determination. "Anything?"

Noah tried to shake his head while maintaining ear contact with the door. There was a small click and the massive doors swung open, slow and silent. Dana and Noah exchanged a glance, and she knew they were thinking the same thing: big doors that open unexpectedly in creepy empty old houses are supposed to *creak,* even if it was in the middle of a sunny spring afternoon and not a dark and moonless night. They walked through the doorway, Dana a little more nervous than she would ever have admitted to anyone except Noah.

The foyer of the house must have taken up half the first floor, and was as grand as Dana would have expected: smooth marble floors, walls hung with large paintings that reached all the way up to the vaulted ceiling and stained glass dome. Directly opposite the entrance was a double staircase with a landing in the middle. It had a thick, long swoop of railing Dana couldn't help but picture sliding down.

Several hallways and doors led away from the foyer, and Dana and Noah paused, unsure if they should go up the stairs, down a hall, pick a door, or wait where they stood.

A voice crackled from an intercom they couldn't see.

"Hello, how can I help you?"

The voice bounced around the marbled chamber and echoed off the glass ceiling, making it difficult to tell where the sound began or ended.

Dana wondered if it was a real person speaking or an electronic assistant — it was hard to guess with this house whether it was run by high-tech robots or old-fashioned servants.

"Hi... I, we're here to pick up forms for the STEM contest," she said loudly, and — not knowing where the voice was coming from — in no direction in particular.

"Down the hall to your left, please," the hidden speaker said tinnily.

They walked down the hall, past staff portraits and a display case full of competition trophies. At the end of the hall was an ordinary door, with an ordinary wooden sign that said:

Dr. Cavor
Head of School

Somewhat relieved that something in this place at last looked like a normal school, Dana opened the door, and they went in.

The disembodied voice from the speaker belonged to a real live person and not a robot. The secretary, an attractive blond, smiled as they entered the office and said, "Dr. Cavor stepped out for a minute but will be back momentarily. You are welcome to have a seat in the inner office while you wait."

Dana checked the name plaque on the secretary's desk, smiled, and said, "Thanks Kevin!"

The nice young man grinned back at her and said, "No problem at all." He gestured towards a dark blue velvet curtain draped across a doorway.

Through the curtain, Dr. Cavor's office was a large octagonal room, with mahogany library shelves on six of the eight walls and stained glass windows on the other two. The vaulted ceiling had a brightly colored skylight dotted with crystals, and the effect of the afternoon sun streaming through panes of colored glass from several directions gave the room an odd, shifting glow, like stepping inside a rainbow. Rainbow-colored shadows and miniature rainbow fragments flitted across the shiny wooden floors.

Dana and Noah had a seat, as Kevin had instructed, sinking down into two pudgy chairs in front of a sprawling but exceptionally tidy desk, built from the same dark wood as the bookshelves. Dana looked around, trying to decide if the slightly swimmy feeling she had was due to the dancing prism rainbows, the enormous desk that seemed more an intricately carved wooden raft than office furniture, the walls of books looming at least two stories high, or the chairs that were like sitting on a bag of black leather pudding.

Near Dana's chair was a mahogany umbrella stand; in it rested a black umbrella so large and dark that it looked less like an umbrella and more like an enormous sleeping bat, its sleek smooth wings cocooned tightly around itself. The twisted black handle protruding from a parachute's worth of silken fabric was easily the sleekest, shiniest item in the room, catching the corners of the skylight crystal's rainbows and reflecting as much light as the umbrella fabric seemed to absorb. To Dana, it almost seemed to glow.

It was formed into a shape — was that a double helix?[4] — that seemed to carefully wind around and through itself. Set in the center of the handle was a somewhat lumpy stone of blue and orange, swirled together like a piece of weirdly-flavored candy. Out of the corner of her eye, the stone seemed to sparkle — or was that still the dancing light from the stained glass window? When she looked at it directly, it had only a lifeless, flat finish.

Curious, she reached out to touch the stone...

Suddenly the walls of books and shiny floor and colored glass ceiling spun around Dana and disappeared. She was no longer sitting in the leather chair but was floating, and the rainbow-filled room had been replaced by a seemingly infinite blackness. Her arms and legs seemed buoyant but also heavy, as if she were frozen solid in a dark lake. In the darkness around her, floating like pool toys, were symbols and images, some bobbing near enough

---

[4] A *double helix* is the shape of a DNA molecule. It looks like a twisted ladder.

to reach out and touch if she could move her arms, some winking in the distance like stars. She had an overall familiar feeling, like finding a box of mementos, or remembering a dream she had had before.

She focused on the object most directly in front of her. It was small and darkly orange, irregularly shaped like a rock, or maybe a very lumpy tangerine, and as she looked at it, she had the overwhelming sensation that she *knew* this, this thing floating in the immense void around her. It began to rotate slowly in the darkness, and as it spun faster and faster, it swirled messy blobs of muted gray and brown around itself, like garbage in a river eddy. The swirls began to take shape, first in a wad around the stone, then another lump beneath that, and as that lump added a long thin mass to each side Dana realized it was starting to look like a human figure, a torso with rag-covered arms stretched out from its baggy body and a faceless head covered in what looked like a garbage hat. Around it, the inky black had shifted too, and now looked more like fog in a moonless sky. It was no longer empty either: dense gray forms were gathering in the mist, like a crowd of darkening storm clouds shaped like people. They seemed to move in unison with the figure, now nearly solid and body-shaped and coming towards her.

Dana wanted to scream. The sound was too thick to escape her throat, and her mouth felt like it was stuffed with straw. She wanted to run, or kick, or do something — anything! — but her body felt as if it were not quite her own. She tried desperately to clench her fists, and she realized that she was grasping the spiraled, shiny handle of the black umbrella. Almost as if by somebody else's control, she raised the umbrella to defend herself against the approaching terrible lumpy garbage body and the shadowy horde it seemed to command. A single glowing green eye blazed to life on its face as the figure reached a raggedy arm-like appendage out to touch her. She thrust the umbrella in front of her and it burst open, unfurling itself as the fog began to swallow her.

As suddenly as she had spun into the dark vision, she was spinning back out of the gray mist.

From the notebook of Dana Dash

CHAPTER 2 — THE DOCTOR IS IN

Dana was back in Dr. Cavor's office. She jerked her hand away from the umbrella as if it had bitten her. "Dana...? Hey, are you ok?" asked Noah.

"Ye... Yeah," she stuttered, her throat now able to make sounds. "Just got a shock there... static electricity..."

Her mind was racing. What just happened? She had been in another place for a minute. She felt the lingering sensation she had when her alarm clock blared to jolt her up for school, when for a few minutes she wasn't sure what was real and what wasn't. But she'd been awake when she sat down, and the squishy chairs were cozy but not so much that she would have dozed off. How much time had passed? Noah was looking at her quizzically, as if he had asked her a question and she had stopped talking mid-sentence.

Dana's heart was thumping in her chest and her blood was pounding in her ears. But she hadn't moved a muscle from the chair in Dr. Cavor's strangely lit office. She shook her head as if to shake away the feeling of immobility, and bounced up and down on the plushy leather cushion just a little, just enough to feel something solid. Gravity — gravity was real. The chair beneath her was real. Noah beside her was real. She unclenched her fist, and saw that her fingernails had dug imprints in the skin of her palm. Could it all have been a dream? It had *felt* real, as real as sitting here right now did. Except for the feeling that she couldn't move, and when she *had* moved it was like being an observer in her own body. Not entirely unlike a dream. What other explanation could there be? Had she passed out? Had some sort of hallucination? A seizure? A sudden nap attack? Had the strange dancing crystal rainbows hypnotized her into an out-of-body experience? How could she figure out what had happened?

Repeat the experiment! The basic tool of the scientific method. She had touched the umbrella, and then... something had happened. *Try it again and see if something happens this time. Two data points are better than one, right?* She started to reach out her hand, still trembling from the adrenaline of the maybe-dream-maybe not-a-dream...

"Are you sure you're ok?" asked Noah again. "You're giving that umbrella a very strange look."

Dana turned to him, not knowing quite what to say. The blue curtain slid velvetly aside and a woman entered the room.

She was an ordinary-sized woman, of ordinary age (probably — Dana wasn't very good at judging how old adults were), with ordinary-looking glasses perched in the middle of her face where ordinary people ordinarily wear them, but the ordinariness about her ended right there at the tip of her ordinary nose.

She was wearing clothes that looked, as near as Dana could guess, to be taken in equal parts from various museum exhibits and the basement of a small-town theater. Dana's mother dressed very sensibly, and Dana was fairly happy to pull on the first clean shirt she found in the morning so she was a little unsure what current fashions were, but she was pretty sure that what this woman was wearing was not it. Which was a shame because, if more people dressed like Victorian space pirates, Dana would not think grown-ups were quite so boring.

"Hello children, and welcome to the Cavor School. As you have certainly gathered from the sign on the door, you are sitting in the office of Dr. Catherine Cavor, and I am she." Dr. Cavor proceeded to walk around to sit behind the very large wooden desk. In contrast to everything else in the room, including Dr. Cavor, the chair was a normal desk chair, the same kind that Principal Peters had in his office and Dana's mother had at her desk at home. Her mother frequently complained about how it hurt her back when she worked for long stretches of time, as she often did. Somehow, Dana could not imagine that Dr. Cavor had ever uttered the phrase "lumbar[5] support."

"Kevin has informed me that you would like to enter this year's STEM contest?" Dr. Cavor leaned back slightly in the plain black chair, and it squeaked a tiny protest.

Except for the chair, no one said anything. Dana usually took the lead, but this time she stared somewhat blankly at Dr. Cavor. Finally Noah said, "Um, yes. We wanted to pick up the rules and entry forms. We're very excited about entering."

"Of course," said Dr. Cavor. She opened a desk drawer and pulled out a folder, leaning forward in her chair to skim its contents. Dana stared at Dr. Cavor's eyes as they darted across pages. The glasses she wore hid her eyes somewhat, as glasses tend to do, but from the side where Dana sat, she could see Dr. Cavor's eyes without the glass obstruction. They were strange, like opals — fiery and bright, but the color seemed to shift, first green, then blue, then gray, then deep black. Dr. Cavor looked up, and the color snapped into focus, an ordinary hazel. She pulled some papers out of the folder she had been examining.

"It is very pleasing to see children who take the initiative to pick up the forms themselves, even if you are quite late in doing so." Dr. Cavor slid the contest forms across the boat of a desk towards the children. Dana was still somewhat mute and immobilized. She was trying to focus on Dr. Cavor's words, but she felt a bit whiplashed at snapping from the floating-space-void-creature-dream (she had decided, until presented with more evidence, to call it a dream) back to the very regular form of a pile of papers.

Noah looked at her, then back to Dr. Cavor. "Ok..." he said, "Then we'll, uh, take these..." He stood and took the papers from the edge of the

---

[5] *Lumbar* describes the lower part of the spine, made up of five vertebrae (bones) L1–L5. It is a common area for people to experience back pain.

desk and stuffed them into his backpack, then looked to Dana, whose mouth was slightly open as if she were about to say something but was not making any actual mouth sounds, so he continued. "Well, we'll be going now. It was nice to meet you."

Dana followed his lead exactly not at all and remained a Dana-esque lump in Dr. Cavor's pudding chair. She tried the mental trick she usually did to clear her head of overwhelming feelings, but it wasn't working. Her mind felt like it had been split into two pieces, and trying to focus on a single thought was like trying to smoosh two stones together. Dr. Cavor gazed through her glasses at Dana, unblinking. Dana returned the look, as if they had telepathically agreed to a staring contest for what seemed like an agonizing length of time but, in all actuality, was likely the non-blink of an eye.

Dana did blink first, and then Dr. Cavor was smiling and opening another desk drawer. "Would you children care for a treat? I am very fond of Unclaimed Babies," she said, pulling a bag of candy out of the drawer. Dana, for the second time that day, would have wondered if her mind was playing tricks on her, except that Noah had not seemed to notice that the spell was broken and was still standing between Dr. Cavor's desk and Dana's chair, tugging at his shirt tails like he did when he was distressed.

Wait, what did she say?

"What are Unclaimed Babies?" Noah asked, continuing to tug at the hem of his shirt. Of all the odd-ness about the Cavor School — the strangeness of the manor house, the office, Dr. Cavor, and pretty much everything that had happened since they set foot on the school grounds — he seemed most unnerved by Dana's complete lack of motion. He had probably never seen her sit this still, for this long. Not ever. She probably never had.

Dr. Cavor opened the bag and shook it a little. "Oh yes. I believe they are now more commonly called Jelly Babies. They are not entirely unlike a Gummy Bear, except it is a baby instead of a bear whose head you bite off. Although I like to start at the feet and work my way up." She took a small red jelly candy from the bag and delicately nibbled at its baby shoes.

It was Noah's turn to stare with his mouth open, as if this act of sweet cannibalism was just one weird thing too many.

Seeing this, Dana snapped out of her stupor. "Yes, thank you very much. Gummy Bears are my favorite." She reached her hand into the bag and pulled forth a small, vaguely-baby-shaped yellow candy. "I think I'll start with the head," she said, and took a bite.

At this, Noah took Dana by her non-candy-baby-holding hand and pulled her gently but very seriously out of the chair, as if to say, *That's quite enough of that, it's time to go right now and I really mean it.* Noah was very good at communicating a lot of feelings in nonverbal gestures. It was a talent Dana greatly admired in him.

"Right... nice to... time to work... nice to meet... I mean, um... thanks," she stammered as Noah pulled her towards the door at a pace that bordered on impolite. She barely managed to grab her backpack from next to her chair.

As Noah pushed aside the velvet curtain and pulled her through, she cast one last look at the umbrella, still resting innocently in the umbrella stand, next to the oddly large desk and the unusual woman, as if it were trying desperately to pretend that it was actually the only ordinary thing in the whole room. Dr. Cavor looked at them as they left, with that same steady, unblinking gaze. Dana couldn't be sure, but behind her glasses the woman's strange gemstone-like eyes seemed to twinkle.

"Goodbye, children!" she called after them. "Oh look, this one has two heads..."

They passed Kevin at his desk, who waved cheerily. Noah, still pulling Dana by the hand, rushed past and barely took a second to smile in return. Dana waved with the hand Noah wasn't yanking on — awkwardly, because it was also full of backpack. They speed-walked down the corridor with the faculty photos and trophy case. Dana feared Noah would break into a run through the lobby, but instead he slowed his pace and dropped Dana's hand as they walked across the marble floors. He pushed open the giant front door, and when it closed silently behind them, he stopped on the manor house porch and looked at Dana.

"Seriously, Dana. Are you ok?" asked Noah.

"Are you?" she replied, a little concerned but also a little huffy. He was the one who dragged her out of there. And refused candy!

"Yeah. Yeah," he said. "Just, I don't know, it got weird in there. Right? What happened to you? You were like a zombie."

"I... I'm not sure," Dana answered, which she realized, now that they were outside the building, was the honest truth. She had touched the umbrella handle, and then...

"I had some sort of vivid waking dream when I touched that old umbrella. It made my head feel really weird." She sat down on the steps. "Maybe I was just nervous, I don't know. There's a lot that's happened since yesterday." She thought of sitting in class this morning, knowing that every student there was whispering about or passing notes about or thinking about the explosion yesterday. She had gotten through the hours by taking deep breaths and reminding herself that as soon as the bell rang, she would be going to pick up the contest forms. The contest was going to save her from the gossip, from the humiliation, from the injustice of being branded a troublemaker and a terrorist-in-the-making. The contest was her best hope for keeping science in her life. "I think I just need a minute."

Noah studied her for a moment, concern in his eyes. "Of course. Whatever you need," he said. He walked down the manor steps and towards the garden he had eyed on their way in. Dana was left alone with her thoughts.

She could already feel the images fading in her mind, the way a dream did after she woke up. What had seemed so real moments ago might be totally gone in a few more. Dana opened her new lab book and started sketching and jotting down notes before the experience slipped away.

What are the possibilities? The past two days had been stressful, but was it really enough to push her into an almost psychotic episode? *What other explanations are there?* Was it down to either insanity or some sort of magic mind powers? She had heard the term "daydream" before, but thought that merely meant idle thoughts. *Is this something other people actually regularly experience?*

The strange garbage-monster-zombie-fog-dream-vision was still lurking on the edges of her mind, like something she remembered remembering. And since she had certainly never been attacked by a floating stone with a trash body in a black void filled with — she thought about all the other floating images, and fished for what the best way to describe them. *Like computer icons, like if I had touched any one of them it would have opened up into an equally weird vision-memory-dream file.* Déjà-vu-space-monster — she shook her head, as if to shake off how ridiculous it all sounded. She didn't like that she had no rational explanation. It was all feelings and vagueness and speculation, nothing that could be readily quantified.

For now, the best thing to do was to suspend judgment until she had more information. Maybe she could get her hands on that umbrella again and test her hypothesis. If something real had happened, she might learn something extraordinary. If the event couldn't be replicated and was all in her head... well, it would be important to know that too. "Experimentation is the collection of data points, and all information is data," her dad had told her. "There's no such thing as failure in experiments. The unexpected result, the rejected hypothesis, yields as much or more information than confirmation of what you already believed."

She sketched the black umbrella — open, as she had seen it in her mind. Looking at it, she thought that this was, by far, the weirdest first page she had ever included in one of her project journals. As she closed the book, she realized that Noah had returned and had probably gotten at least a quick glimpse of the page over her shoulder.

"Ready to go?" he asked.

"Yeah. Let's head over to my house."

"Sounds good," he said, but she detected a certain worried tone in his voice.

From the notebook of Catherine Cavor

CHAPTER 3 — DASH HOME

Nothing helped Dana clear her head more than moving fast. She finished ahead of Noah in the race back to her house, but just barely. "I win again. Dash by name..."

"Yeah, yeah... dash by nature. I've heard that one already," said Noah. "But I almost had you that time. I was ahead..."

"Keep trying, my friend. They don't call me 'Dash the Flash' for nothing!"

"Actually, they don't call you that. No one ever has."

"Well, if they did, it certainly wouldn't be for nothing."

Noah remained quiet, perhaps unsure if it was even possible to disagree with that statement.

"I'm home! Noah's with me!" Dana yelled, as they came through the front door, throwing her helmet on the entryway bench.

A muffled reply came from the direction of the basement. Her father was, as usual, downstairs in the workshop.

"Hi Mr. Dash!" Noah shouted back.

Noah's voice was answered by rapid-fire squeal-barks accompanied by an escalating thunder of animal footfalls. A large brown dog slid excitedly into the entryway like a cartoon, his thick tail whacking back and forth against walls and doorways and any part of human or house that got in its way. "Ouch!" said Noah, "Take it easy, Bosco!" Hearing his name, Bosco rose onto his back legs, put his paws on Noah's shoulders, and licked his face, as his powerful tail continued to motor at the front hall shoe cubbies *wumpwumpwumpwump*.

"This dog," laughed Noah, pushing Bosco off his shoulders and trying now to dodge both tongue and very heavy tail, "was obviously a spectacular failure of some sort of breeding program attempting to bioengineer a canine with weapons on both ends."

Dana laughed too. "Well, the front end doesn't look very threatening." She pointed to Bosco's big pink tongue lolling dopily out of the side of his mouth that, while containing the usual number of sharp canine teeth, looked to be smiling happily as Noah scratched him behind one ear.

Satisfied with a good few minutes of ear-scratching and *who's-a-good-boy*-ing, Bosco allowed them to proceed through the living room and kitchen, and downstairs into the basement, padding contentedly after them.

They found the door to the workshop closed and Dana knocked lightly. "Dad? Can we come in?"

"Hiya Bug! Don't let Bosco in, I'm cleaning Chip."

"Ok," she said. "We'll put him outside. Come on boy!"

Dana walked with Noah and Bosco through the downstairs game room and out the back door onto the deck.

"What's a chip?" Noah asked.

"Not what, who. It's an acronym. The first letters of each word spell CHIP; it stands for Closed Head Injury Patient." Dana took Bosco's white Frisbee from the deck box and his tail motor whirred into action; it *wump-wumpwumped* against the deck boards and he squeal-barked until she opened the back gate. He bolted down the deck stairs.

"Chip is a red fox who was running around the neighborhood this winter until a car hit him. Mom found him in front of our house, and Dad built a cage for him in the workshop to try and nurse him back to health. They talked about calling animal control, but Dad was afraid they would put him down. I don't know how long we'll be able to keep him, since he is a wild animal and doesn't belong in a cage. But he's started to feel like part of the family."

Dana looked down at the bottom of the yard, where Bosco was squeal-barking. "Bosco, fetch!" she shouted. She threw the Frisbee over the deck railing, over the terraced vegetable garden with its chicken wire cages to keep out the deer and other critters, past the terrace lawn with the picnic table, and into the woods. Bosco charged after it like he had been given a very serious order to seek and destroy.

"Cool. Can I see this fox?" asked Noah.

"In a minute — Bosco needs a little exercise. He loves the chase, but he really doesn't like being on a leash." The dog paraded out of the woods, back up the hill and the stairs, head high, eyes gleaming, Frisbee proudly displayed in his mouth, a few twigs and leaves stuck to his head like trophies of a hard-fought battle.

Bosco dropped the Frisbee at her feet. Dana bent down, picked it up, and flipped it over the railing again.

"But he certainly makes up for lost walks with the exercise he gets running up and down this cliff," she said.

"Yeah," Noah said, "A lot of Labradors get fat, but these cliffs seem to keep him in good shape. How old is he?"

"We're not sure exactly, but he's at least twelve. That's old for a Lab, but he still acts like a puppy."

Dana continued throwing the Frisbee for Bosco while Noah sat down on one of the deck chairs and pulled the STEM contest paperwork out of his backpack. After a few minutes, during which Bosco was several times praised as a "Good Boy" for bringing back the Frisbee for Dana to throw it once again, Noah looked up from the forms and spoke.

"It really does look like a good idea to team up," he said. "In previous years, the team would have to split a scholarship. But this year, we get the benefit of collaboration and we each get a full scholarship of our own. It's a good way to reward scientific cooperation."

Dana stared over the deck railing, watching the Hudson River[6] in the distance. Henry Hudson — his quest had ended tragically, her mother had told her, when his ship became stranded, held captive by the winter ice of the immense inland northern sea he called Mare Magnum, now named

---

[6] Technically the Hudson River is not a river but a tidal estuary, which means that the lower half of the river has a tide like the ocean. It was called the Muhheakantuck — "river that flows two ways" — by the Mohicans, and later named after explorer Henry Hudson (1565-1611), who sailed it on his ship the *Half Moon*, looking for the legendary Northwest Passage.

Hudson Bay. Dana's mother told of the hubris of the explorer, his refusal to listen to his fellow sailors as they begged him to turn back, their mutiny, and how he was set adrift in a small craft on the dark, cold waters with his son and a few loyal men, never to be heard from again.

Dana thought about her own inclination to control everything, to do it all on her own. "I'd certainly rather have you on my team than be working against you," she said, still gazing at the river as it flowed inland with the incoming tide. "And I want to be with you at school next year, wherever we are. Either we both win or we don't, together. We just need to come up with an idea we both like. We both like science stuff, but we have very different interests."

Noah got up and joined her at the deck railing. "I'm sure we can figure something out." He smiled, then noticed something on the deck railing. "Hey little guy, what are you?" he said, bending to look at a tiny lizard, almost the same color as the deck, trying to blend into the wood where the railing met the post.

It froze for a second as Noah inched closer, as if realizing its cover had been blown and pondering its options. It seemed to decide that between fight and flight, it was unlikely to win a battle against the fierce opponent that was a ten-year-old boy, and it hurried along the railing in the exact opposite direction of Noah. Unfortunately for the lizard, this meant running directly past a ten-year-old girl. Dana's hand shot out. She deftly scooped the lizard into one hand, cupping her other hand over it. Noah raised his eyebrows, impressed, though not especially surprised.

"Ah, we see these a lot. They are really tricky to see — it looked brown on the railing but they are actually kinda transparent, so they blend into their surroundings. Look."

She held her cupped hands towards Noah, moving one thumb slightly for Noah to peek into the lizard's makeshift cage. In the shadow of her hand, the lizard's body had a glowy reddish hue, like it was directly on top of a flashlight.

"You can see them better at night. Then they're pale translucent white, almost glow-in-the-dark."

"I've never seen anything like it," said Noah, "It looks like an anole — a small chameleon from down south. But you never see them this far north, and it's also way too small. Is this full grown?"

"I dunno, you're the biology expert, not me. That's as big as I've ever seen them. You really never see these at your house?"

Noah shook his head. "Nope. Not at my house. Or on any of the nature walks I've done around Dash-on-Hudson. Or anywhere else, either."

"Well, they certainly like hanging out at our house."

"Can I have that one? I'd like to take him home and see if I can do a little more research."

Bosco came back up the stairs, but slower this time, and instead of returning the Frisbee, he laid it on the deck and then plopped on top of it.

"Ok," said Dana. "Promise not to hurt him, and that you'll let him go when you're done." Dana looked over at Bosco, panting but contented on the deck, the Frisbee hidden from sight. "Speaking of done, Bosco is. Let's go inside; Mr. Mystery Lizard can stay out here for now." She turned a small empty flower pot over on top of the lizard, trapping it on the surface of the deck table. "Don't forget to come get it later."

From the notebook of Noah Knight

They left Bosco on the deck, contentedly gnawing on the white disk once he was certain the throw-things-off-the-cliff-into-the-woods game was over.[7]

Dana knocked on the workshop door again. "Bosco's outside. Can we come in?"

"Come on in. I'm just finishing up," Dana's father said.

Dana pulled the door open, and they entered the converted garage workshop. Their house was old and had been in her family since the Dash family had founded Dash-on-Hudson. Dana's father had converted the garage himself and had painted the inside of the garage door and the floor like a checkerboard. Shelves with computer and science books hung above work tables covered with computer and quadcopter drone parts. One wall contained drawings and sketches, some technical-looking, some cartoonish.

In front of the work tables, set like game pieces on the checkerboard floor, were two short battered gray file cabinets. On top of one was a medium-sized cage. On top of the other was a frame and harness contraption that was holding the injured young fox mostly immobile as Dana's father tended his injuries.

Mr. Dash said, "Noah, meet Chip. Don't move quickly or make any loud noises, but he's a lot less skittish than he was in the beginning, so if you want to touch him you can. Let me finish these bandages first."

He was tying off strips of white gauze behind the fox's head. The bandages were wound around the fox's skull in a way that kept the left eye, both ears, and the muzzle uncovered. His hands worked quickly as his hazel eyes peered over small rimmed glasses.

"So what are you young geniuses up to today?" he asked as he adjusted the bandages.

Dana and Noah both started to speak at the same time but Noah stopped and let Dana continue, "The science fair at the Cavor School. We got the forms!"

---

[7] Chasing is a fun game, but chewing is a good one too.

"Right, right," said Mr. Dash. He cut the last excess material from the bandages and stepped away from the table. "Ok, you can pet him now. Stay away from his mouth. He hasn't tried to bite me in a while but he's still a wild animal."

Noah looked at Dana, who nodded and said, "Go ahead."

Noah walked slowly towards the table with his hand outstretched, and lightly touched the fox's back. He could feel the fox shaking a little as he gently stroked its red-orange fur. "Easy there, Chip," Noah cooed, "It's okay. The Dash family is taking good care of you." He stepped back and asked, "How long is his recovery going to be? Are you going to release him soon?"

Mr. Dash unhooked the harness. "I'm not sure. The wounds are healing well. But he's lost one eye and his behavior is still erratic. He may not ever be able to survive on his own in the wild." He lifted the fox and crossed to the cage, placing it gently inside. "He keeps rubbing his head on the cage and reopening his wounds. I also had to fill the bottom of his water bowl with rocks to weigh it down. He wouldn't stop knocking it over and trying to dig under it."

Mr. Dash replaced the cage top, and then took off the gloves and returned them to one of the drawers. "Someone's car fender clonked him pretty hard on the head and I'm afraid there's probably permanent brain damage. He's lucky to be alive but I'm not sure how great his life is going to be. I should probably find him one of those animal sanctuaries that takes in wounded wild creatures, but..." He gazed at the fox, who had curled up in a circle in his cage. Then Mr. Dash turned to them and said, "Anyway, good to see you, Noah."

He extended his right hand and Noah shook it. Even when Dana's father smiled, as he did now, Dana could always see that the smile never quite reached his eyes — eyes that were so much like her own except they always looked somewhat sad. He had once worked in computer science, before Dana was born. Now he spent his time at home, taking care of Dana and her mother, fixing up the house and puttering around with an array of inventions and projects. Fixing a broken fox was one of the latest.

"Nice to see you too, Mr. Dash," said Noah, in the serious tone he reserved for interacting with adults. He noticed a silver medical bracelet on Mr. Dash's wrist. "Are you allergic to something?"

"Oh yes," said Mr. Dash. "I'm highly allergic to death."

From the notebook of Donald Dash

CHAPTER 4 — COOL IDEAS

Noah's forehead wrinkled in confusion. "What do you mean, you're allergic to death?"

Dana interjected, "He's baiting you into discussing his religion. That's a cryonics bracelet he's wearing."

"Cryonics?" asked Noah.

"It's not a religion, exactly," said Mr. Dash, "but it is a shot at an afterlife. The bracelet tells emergency medical personnel that my body will be donated to science. I'm signed up to be frozen in liquid nitrogen when I die. At near absolute zero,[8] the information in my brain will be preserved. If some way to save me is ever invented, it would be like I died and came back to life."

"Wild!" said Noah. "Does it cost a lot?"

"It's not free, but funerals are expensive too — it's well within most people's means. You buy a life insurance policy..." He trailed off, noticing Dana's dramatic eye-roll. "Well, if you're interested, you can look it up..." he trailed off again.

"So: the Cavor School STEM contest. Dana, you may not remember, but you've met Dr. Cavor before. You attended daycare at that school. Maybe that will help you get chosen."

Dana said, "Well, first we need an idea for our project proposal."

"Well, let me know if you mad scientists need any help plotting your evil schemes. Otherwise, let me know when you're ready for dinner." He ruffled Dana's already messy hair and left them there in the workshop.

[8] *Absolute zero* is -273°C or -460°F, which is as cold as you can get. At this temperature, molecules stop moving.

An hour later, Dana and Noah were still discussing ideas. Dana wanted to do a mechanical project — she was in the mood to build a cool machine. Noah, however, was more interested in biology; he kept coming back to projects involving plants and animals.

"Look," said Dana, "Maybe we should do separate projects. I definitely want to use these computer-controlled servos I've been playing with." She held up a green control board and a small electric motor. "Robot control systems are cool. Coding is cool. Arc welding is cool. I want something that moves. Maybe a rocket? And I'm sorry, but I don't want to grow new kinds of mold when we could be the first ten-year-olds to actually put something into orbit!" She was getting frustrated.

"Genetics is cool too," insisted Noah. "Biohacking is cutting edge. You can build brand new organisms right in your garage!"

Dana made an effort to calm herself. "Maybe we should take a break. Want to play a game? Clue, like we used to? You can be Professor Plum."

"Ha! Right. I was always Professor Plum. I think you even called me that for a whole month in third grade. But don't you prefer Monopoly?"

"Yeah. I'd eat pretzel rings and put one in my eye as a monocle to look like the guy on the box."

"Oooh! Pretzels sound good!"

"Snack break!" said Dana. "Be back in a jiffy." She ran upstairs to grab some pretzels and found a jar of Gif[9] peanut butter to dip them in.

She slowly walked back down to the workshop, still thinking about potential projects.

Noah wasn't there, but he came in behind her. "I went back out to the deck to look at that strange little lizard again," he said, "but it had escaped somehow."

"Oh yeah, they do that. I've caught them before and they're slippery little buggers."

---

[9] Many people are sure they remember this peanut butter being called one name, even though it has always been another. This is an example of the *Mandela Effect*. Look it up!

"But how could it? It can't be strong enough to lift that flower pot."

"I don't know," said Dana. "I had one in a glass jar with some dirt and grass once. In the morning the lid was still on but there was only the dirt and grass."

"Weird," said Noah. He examined black smudges on his fingertips and then rubbed them on his shirt. "So, Monopoly?"

"How about something new?" asked Dana. "I have a starter set of Elementalist cards."

"Yeah!" said Noah. "I've heard of that. It's a card game based on that cartoon where people summon monsters to battle each other?"

"Sort of. It's set in a world where people can summon nature spirits, called Elementals. But instead of just Earth, Air, Fire, and Water, they can summon the whole periodic table."

"Oh, so there are a hundred and something of them, like the real periodic table of elements."

"Right. But even more than that, because different isotopes of the same element can also be summoned and have different abilities. Some are magnetic, or radioactive, whatever, which gives those Elementals special attack powers."

Dana got the game and they read through the rules quickly, having a good laugh at the part about the Great Wizard Mendeleev,[10] and soon they were playing a test duel between two Elementals for control of a new pocket universe.

Noah had an initial advantage. He started with a period 4 Chromium Elemental, and he managed to step up to a period 6 Tungsten Elemental named "Wolfram." Dana had started with period 2 Carbon elemental but was only able to step it up to a period 5 Tin elemental named "Stanus." Noah's larger Elemental might have won, but Dana managed to cleverly

---

[10] Dimitri Ivanavich Mendeleev (1834-1907) was a Russian chemist who first noticed that the chemical properties of elements repeated in a periodic fashion as the atomic masses increased. He created the first rudimentary version of the periodic table widely in use today.

keep the temperature of the pocket universe in a range where Stanus could choose to be either solid or liquid. This was enough of an advantage to allow Dana to finally eke out a victory.

"I win!" Dana yelled triumphantly.

Noah sighed, then got up and wandered over to look at the drawings on the wall. "What is this?" he asked, pointing to a child's drawing of a rainbow-colored dog-like thing.

"Oh, that's ABC." Dana replied.

"ABC?"

"Yeah, Alien Bralien Chipalien. He was my imaginary alien fox friend. I don't remember it but my parents tell me I used to tell stories about him when I was little. Probably why Dad called his rescue fox Chip."

"And this drawing here?" Noah pointed to a much better cartoon drawing of some people standing on a tiny moon. "That's your family right? Bosco chasing his Frisbee in space. And that's your mom holding Chip."

"That's our Newtonmas card last year. Funny thing though — we didn't actually have Chip yet. My mom asked one of her artist friends to do a Christmas card based on the Little Prince story, with all of us on the little asteroid. The artist added the fox, because the Little Prince story has a magical fox in it. Then a couple months later, presto, we have a fox. Maybe, subconsciously, these pictures are why my dad decided to keep Chip."

"Huh. Weird coincidence," said Noah. Then he asked, "What's Newtonmas?"

"Oh, that's what Dad calls Christmas. He likes to rename holidays for famous scientists who were born or died on that day. Sir Isaac Newton, who gave us a theory of gravity, and the equations to plot astronomical orbits, was born on December twenty-fifth. I was born on January first, and Dad says if I do something as impressive as Sir Isaac, someday he can change the name of New Year's Day to Dashmas."

Noah laughed and looked back at Chip in his cage.

Dana continued, "Unfortunately I may be up against Grace Hopper on that. She was one of the first developers of a high-level computer language, working on the UNIVAC I.[11] She died on January first. Although my dad does try to match birthdays first before looking at deathdays."

Just then Chip started and jumped at the side of the cage, banging his head hard against the bars.

"Ow," said Noah sympathetically, "That looked like it hurt. Maybe he needs a bigger cage or something." Chip started digging and biting at his water dish. He splashed a little bit but the rocks in the bottom weighed it down and kept it from turning over.

"Dad said that animals can be ok in small spaces as long as their basic needs are met. Zoo animals that normally have more than one-hundred-square-kilometer[12] territories can survive in habitats smaller than one hundred square meters.[13]" Dana came over to Noah and they stood looking at the fox. "Mom says that goes for people too. She works in Manhattan."

"Hmm," said Noah, "I wonder how much space human animals really need to stay alive?"

---

[11] *UNIVAC* stands for "Universal Automatic Computer." The UNIVAC I was one of the first widely available mainframe computers.

[12] A *kilometer* is about half a mile. (1km ≈ 0.62 miles)

[13] A *meter* is a little bit longer than a yard. (1m ≈ 3.28 feet)

"Well, space capsules need to be super small and lightweight, but the crews have to survive long journeys in them."

Noah thought about this. "A space capsule is a totally closed system — a separate biosphere — with its waste and air recycled. Maybe genetically engineered algae could produce oxygen?"

Dana nodded and said, "And computer-controlled systems for the oxygen levels."

Noah continued, the excitement in his voice rising. "Controls for gas levels in the algae tanks too," he said, "and sensors and servo-controlled shutters to keep the sunlight levels right for the algae. And the whole biosphere would need to be self-contained — fully enclosed. Maybe a metal frame with plexiglass windows?"

Dana grinned and said, "You know what, Professor Plum, it sounds like we might even need to do some arc welding."

Noah grinned back at her. They had at last found a project.

It was dark when Dana walked Noah outside and said goodbye on the front yard where his mother Min was waiting in their small car. Dana waved as they drove away down the street, still filled with excitement at the thought of building their own spacecraft.

Dana was about to step back inside when, out of the corner of her eye, she saw something move. Across the street, at the very edge of the yellow glow of the streetlight, she thought she saw a figure nestled in the overgrown bushes. She wasn't sure what she was seeing. It was the same feeling when she woke up in the middle of the night and there was a strange shape in the corner of her bedroom — she knew rationally that it was nothing, probably a lump of dirty clothes on a chair, but some primitive part of her brain screamed *DANGER!* and she would freeze, heart racing, ready to run or fight.

Dana stared into the darkness. The more she peered into the shadows, the more she saw a face looking back at her.

CHAPTER 5 — UNDER PRESSURE

Part of her wanted to leap inside the safety of the house and slam the door behind her, but she resisted her fears and slowly bent down, pretending to tie her shoe. Keeping an eye on the form in the shadows, she inched her hand to the ground and felt for one of the small stones lining the walkway. She felt a strange buzzing pressure in her head as she rose and threw the rock, as fast and hard as she could.

The white stone arced through the light of the lamp, and there was a rustle in the leafy shadows. Maybe a flash of green? The hooded figure from her strange dream-vision in Dr. Cavor's office flashed through her mind. The stone thumped in the leaves, hitting something solid. Then silence, as the pressure behind her eyes faded. Dana stared at the spot where she thought something had been. Nothing there but bushes. Had she imagined it?

With a sudden crash, something came barreling out of the hedge into the street. Dana leapt back, even as her mind identified the thing as a large male deer. The tension drained out of Dana's body and she released a lungful of air that she hadn't realized she was holding in.

Perfectly normal. Plenty of wildlife came out of the nearby woods and roamed the streets of Dash-on-Hudson at night, especially during the winter and spring when food was scarce. She watched the big buck, its hooves scraping softly against the asphalt with each footfall, and noticed something on its head and neck. *Almost like there's a smaller animal riding the deer. Or maybe it's got some trash caught on its antlers?* Before she could really process what she was seeing, the deer had disappeared from the ring of streetlight and back into the night.

With her heart still pounding like she had been racing against Noah, Dana turned and went in the house, shaking her head. She really needed to understand whatever had happened to her in Dr. Cavor's office, if only so she stopped jumping at shadows. But right now, she needed to get some sleep.

In the morning, the intensity of the vision in Dr. Cavor's office had almost vanished. When Dana opened up her notebook to go over her project notes, she looked at the picture she had drawn of the umbrella. Why had she gotten so worked up over a silly daydream that she now could barely even remember? It was certainly unscientific to have this in her project notebook. She almost tore it out, but she stopped herself.[14]

Instead, Dana found a paperclip and attached the first two pages of the notebook together, so the umbrella picture wouldn't distract her. She then read over the project notes from her discussion with Noah. They had descriptions of computer-controlled shades to manage sunlight, extra grow lights to keep the algae happy at night, and monitoring systems for oxygen and carbon dioxide levels. She had made a first concept sketch of their biosphere. It was a ball on legs that looked a bit like a moon lander.

From the notebook of Dana Dash

[14] It's good scientific protocol to never remove pages from a lab book. This is a result of lawsuits in which lawyers have claimed that the suspiciously missing lab book pages were the ones that could have proved their side of the case.

With only two days to write their proposal, they spent the weekend at Dana's computer hashing out the features and details.

"So," said Noah, "We want to be able to demonstrate that our self-contained biosphere could keep a spacecraft crew alive on a real mission. If you're seriously set against using animals as test subjects..."

"It's cruel," said Dana. "I won't do it."

"It's part of science," said Noah. "Biosciences would never have advanced this far without animal testing. It's a small sacrifice to — "

"I said I won't do it," Dana repeated firmly.

"Ok," said Noah, "then the only way to really prove it works is to test it on ourselves. We could set it up a week in advance of the judging and then move in — camp out for a week inside."

Dana considered this. "It could really impress the judges. I love the idea of sleeping out under the stars, plus the whole mad-scientists-experimenting-on-ourselves thing, but how do we prove that it's really working and that we're not cheating? How do we show no outside air is getting in?"

"What if we pressurize it?" asked Noah. "Can we have a pressure sensor and show that the inside of the biosphere is always at a higher pressure than outside?"

"Hmmm," Dana said. "Maybe. It would certainly prove that we were airtight. Or that we could only be losing air out, not getting fresh air coming in."

"How hard would it be?"

"Not very. But it needs to be done right in order to be safe." Dana had an unusually serious tone to her voice. "Not only do we have to make sure your algae system works so we don't run out of air, but we also have to be careful not to blow ourselves up."

"You're joking about that, right? Could it seriously blow up? It's just air pressure."

"Air pressure is no joke. I almost got suspended and put on a terrorist watch list two days ago for an explosion was 'just air pressure.' It's easy to ignore: air is invisible, it's all around us, and we walk through it and hardly feel it. But air pressure is really really strong. Each square centimeter[15] of your skin has about ten newtons[16] of pressure on it all the time. That's more than a hundred thousand newtons on your whole body." Dana always tried to use metric units like "newtons" because they were considered more scientific. She had seen it frustrate Noah, but this time he seemed to know what she was talking about.

"Seriously?" said Noah. "Wouldn't that squash me flat, like being hit by a truck?"

"The forces are 'hit by a truck' level, yeah. But you have pressure inside your body from fluids, and air pressure fills your lungs, so your body is always pushing back from inside equally as hard. There's a whole lot of hidden pressure and potential energy that only becomes obvious when it becomes unbalanced."

"That still seems like way too much force. Are you sure you have that right?"

"I know, it's not intuitive," said Dana. She paused for a moment, and then said, "I think I can show you what I'm talking about. I've been meaning to see this experiment through properly anyway."

Noah followed Dana into the garage. She walked over to a medium-sized steel barrel sitting against the wall, tipped it on its side, and rolled it back to where Noah was standing in the middle of the checkerboard garage floor. Then she grabbed an extra large Bunsen[17] burner stand with a small propane tank from a table and set it on the floor.

---

[15] Between one-third and one-half of an inch.

[16] Named after Sir Isaac Newton, the *newton* is the metric unit of force. If you're more familiar with the British system of weights and measures (as was Sir Isaac), a newton is equal to 0.225 (a bit less than one-quarter) pounds of force. Air pressure at sea level is about 15 pounds per square inch and the air pressure pushing on your whole body is likely greater than 25,000 pounds.

[17] Robert Bunsen was a chemist who did pioneering work in photochemistry and greatly improved the function of the standard laboratory burners that now bear his name.

Noah said, "My grandparents' neighbor tried to deep-fry a turkey on a setup like this. It went spectacularly wrong. Burned the house down on Thanksgiving. Or I guess technically, the house burned up.[18]"

"Well, we won't be using any flammable oil for this trick, so we should be ok... but let's look around for... Good! There's the fire extinguisher. Any experiment should be fun and educational, but also reasonably safe."

"Safety third![19]" Noah laughed.

"Safety third!" Dana grinned and laughed along with him. This was probably her favorite thing about Noah. He was thoughtful and cautious, much more than she was usually, but still adventurous enough to be game for most of Dana's ideas.

Noah adjusted the Bunsen burner so that it was in the center of a checker. He liked symmetry and order, even when it might be followed by an explosion. Together they lifted the metal barrel onto the heating stand, and then Dana removed the small metal cap on the top of the barrel. She poured in some water from a half-filled liter bottle on the table, and put the barrel cap in her pocket — not that she didn't trust Noah, but this time she would know exactly where it was at all times. After handing Noah a pair of safety glasses and putting on her own, she lit the burner.

"At a higher temperature, the same amount of air will have a higher pressure," said Dana. "When that water starts to boil, we'll see the steam coming out of the top of the barrel, and we'll know that the air inside is at one hundred degrees Celsius.[20] Higher temperature means the molecules are bouncing around faster. If some idiot put the cap back on, the hot molecules would bounce around until the barrel explodes. But since

---

[18] True story.

[19] Ok, but at least third, children. Probably second. It should always be high on the list, but admittedly, if safety is always your first concern, you will never actually do anything. Leaving your house isn't completely safe. Lying in bed all day is safer than getting up.

[20] Water boils at 100 degrees Celsius. Really hot! Also known as 212 degrees Fahrenheit but that's harder to remember.

explosions are dangerous,[21] I left the cap off, and instead of the pressure building inside, those atoms of hot air will leave the barrel as steam. So the pressure will be the same, but the barrel will have less air in it."

She opened her notebook and jotted down an equation, saying, "According to the Ideal Gas Law,[22] let's see... have to measure in degrees Kelvin[23] to start at absolute zero... there's about a twenty-five percent increase... of course I'm not accounting for the steam displacing the air, which will make it even more spectacular..."

Soon the water could be heard at a rolling boil inside the barrel, with steam whistling out of the top. Dana turned off the burner, and, putting on a protective glove so as not to get burned, she carefully tightened the metal cap, sealing the barrel. Then she stepped back to stand with Noah.

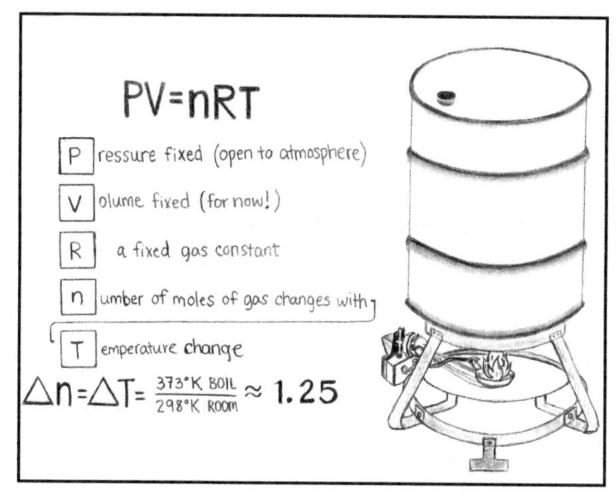

From the notebook of Dana Dash

[21] They really are.

[22] The *Ideal Gas Law* shows the relationship between Pressure, Volume, number of atoms, and Temperature for any gas. It is written PV=nRT. The constant R is based on chosen units of measurement.

[23] The *Kelvin* temperature scale (William Thomas, 1st Baron Kelvin, 1824–1907) has the same size degrees as Celsius, but starts at absolute zero.

"So," Noah said, "there's a lot less air in the barrel than there was before. Now what?"

"Now we wait for the temperature to come back down. And we get to see a demonstration of how much force is really always all around us. Oh! Hang on, I know how we can make this happen faster."

Dana ran out to the big freezer in the basement and returned with a bowl filled with ice cubes. She poured the ice on top of the hot barrel, and the ice began to hiss and steam as it melted. The barrel began to groan and ping. Dana backed away hastily, suddenly remembering those same sounds from her recent disastrous escapade with air pressure.

Noah, looking alarmed by the noises, said, "But you say this won't explode, right?"

"Yes. I mean no. Just the opposite. When it cools down enough there will be — "

BLAM!

It sounded like a gunshot echoing all around the garage walls. The checkerboard garage door shook and the windows rattled, and both of the children jumped.

The steel barrel fell off the burner stand and crashed to the floor, rocking back and forth on its side. It was still metal, but it really wasn't much of a barrel anymore at all. It looked like it had been hit by several large trucks. The sides were crumpled in from three different directions, crushed together in the middle like it had been cinched tightly with a belt. It looked like a giant hand had crushed it as easily as a paper cup.

"...an implosion," Dana finished saying, after she caught her breath from the excitement. "Not as dangerous as an explosion, but certainly no less dramatic. When I put the cap back on, the number of atoms of gas in the barrel was held in place, so something else had to change in our equation as the temperature was lowered. The steel barrel held out as long as it could, but then the equation rebalanced by reducing the volume of the container."

"Wow," breathed Noah, looking at the poor pummeled former barrel. "Ok. I get it. Air pressure is no joke. Do you want to give up on pressurizing the biosphere?"

"No," said Dana, "I'm not giving up on anything. I really like the idea of proving that our biosphere won't have air going into it — that we can actually live for a week making our own air, like if we were really in space. I think that will knock the pants off the contest judges.

"But we will need to build a structure strong enough to withstand significant internal pressure, in case we make a mistake and the pressure builds up. I certainly don't want to be in the middle of an explosion again."

CHAPTER 6 — CRIMES AND MISDEMEANORS

Noah gasped. "Yikes! I'm going to have to trust you on this. I'm working on our air supply system. Building the biosphere structure, and maintaining the right pressure that keeps us from either imploding or exploding, is up to you."

"No problem," said Dana. "But I'll make sure that you understand what I'm doing, so you won't have any doubts. Speaking of which, it's also important for me to believe we are not going to run out of air. Explain again how having a fishtank full of bacteria will give us oxygen to breathe?"

"Not only provide oxygen," Noah said. "It also has to get rid of all the carbon dioxide we breathe out. Removing the CO_2 is as important as making more O_2. Too much carbon dioxide is even worse than not enough oxygen. You can die from it even when you have plenty of oxygen in the air."

"Uh, not building my confidence here, partner..." said Dana. "Please tell me that you have the necessary magic to make all this work."

"Well it's not exactly 'wingardium leviosa,'" said Noah.

"It certainly had better not be," said Dana, "since the transfiguring spell is actually 'vera verto.'"

"Perry Potter nerd!" said Noah affectionately.

"Out and proud," said Dana. "Of course, I prefer Geranium Granger. Now be the wonderful bio-science nerd I know you are and tell me how you work this spell."

"Well I'm starting with a cyanobacteria that already consumes carbon dioxide and excretes oxygen," Noah began. "I need to increase its efficiency so our setup can be reasonably small."

Danna nodded as if of course she knew what cyanobacteria were,[24] and Noah continued.

"So I'm going to do a bit of genetic engineering using a CRISPR-Cas9[25] genome editing kit. I'll be splicing in genes to increase our bacteria's metabolic rate."

"And this CRISPR thing is something you can order online? From MadScientistsRUs?"

"You can indeed! It's a very new technology, but it's really taken off. It uses built-in mechanisms in bacteria that copy and paste DNA in order to 'remember' genetic codes of attacking viruses. But it has been refined to allow very general editing of any genome — even human DNA."

"Wow! That is really cool. Who invented this stuff?" asked Dana.

"Ha! That depends on who you ask," said Noah. "According to the scientific community, it was two women working together, Dr. Duadna and Dr. Charpentier. They've been jointly named as recipients on almost all of the scientific awards in this field."

"But?" asked Dana.

"Well, the United States Patent Office gave the major patents for this technology to a man named Zhang, even though he filed patents later than Duadna and Charpentier."

Dana sat thinking about this for a moment. "Really doesn't seem fair, does it?"

"No," he said. "It doesn't seem fair. It's also not uncommon. A similar thing happened with the discovery of DNA structures — "

Dana stopped him from launching into a bio-science lesson and said, "Ok, back to our project. We'll be camping inside. What if something on

---

[24] Cyanobacteria are *blah blah something blue-green algae*, that's what Dana knew about cyanobacteria. She was very glad that Noah knew so much biology.

[25] *CRISPR* stands for "Clustered Regularly Interspaced Short Palindromic Repeats." Say that ten times fast! *Cas9* is "CRISPR associated protein number 9." Very cool and complicated stuff! But until you find time to study it seriously (and I have no doubt you will want to) just remember CRISPR-Cas9 as the *wingardium leviosa* of genome editing.

the outside needs fixing?"

Noah thought about this and said, "Well, I guess only one of us could stay inside, or we could use lab animals."

"Stop it with the lab animals thing. You know my feelings about that. And we're partners, so we're in this together. Maybe an airlock, so we can come and go without letting in extra air?"

"Wouldn't that be cheating if we can pop outside for fresh air if things start to get stale?" asked Noah.

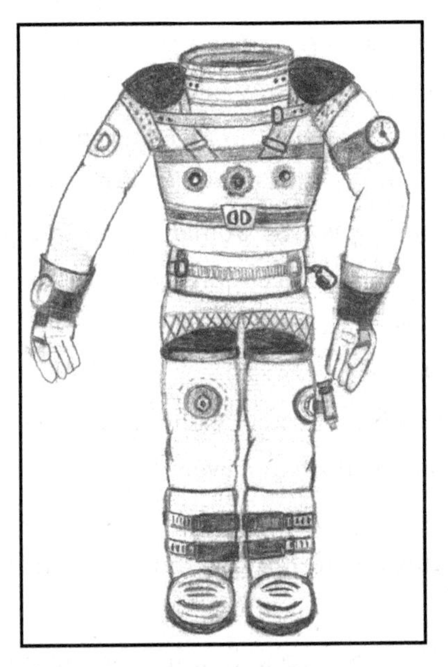

From the notebook of Dana Dash

They thought about this for a little while, then Dana smiled. "I've got it!" she said. "And it's another cool feature! We can build a working spacesuit. Then we can take an EVA[26] while still only breathing our project allotment of air."

"Oooh!" said Noah. "Having a space suit would be really cool. Although

---

[26] *Extra-vehicular activity (EVA)* is any activity done by an astronaut or cosmonaut outside a spacecraft beyond the Earth's appreciable atmosphere.

it sounds like a lot more work. We only have a month at the school to put it all together."

But Dana didn't hear him. She was already sketching plans for a spacesuit.

Friday morning, Dana got up early to print the final version of the project proposal. While it was printing, she got dressed, putting on her good luck T-shirt. Believing in luck might not be very scientific, but her shirt was: it had a picture of Albert Einstein on the front, looking serious, and another picture of him on the back with his tongue sticking out, looking silly. She put the document in a fat manila envelope, addressed it to the Cavor School, and stuck it in her backpack. If this was her last hope to study science, she wasn't taking a chance on the mail. She and Noah would deliver it themselves.

After school, Dana and Noah took a path through the woods separating their school from the Cavor School. The forest ended abruptly at the southern edge of the nearly 250-hectare[27] Cavor School grounds. But before they had gotten very far, Dana spied a lonely little building at the edge of the trees, too large to be a maintenance shed, too small to be a school building, and the wrong shape for a house. The windows were shuttered, and the weeds surrounding it looked as if somebody had forgotten, or wanted to forget, it was there. On impulse she turned towards it. Noah followed.

The weather beaten wooden sign above the door read, "Moonbeam Daycare and Preschool." On the door itself was a dusty, hand-printed sign saying "KEEP OUT."

[27] The metric hectare is a square area of 100 meters each side, or 10,000 square meters. Translating to the quaint old British system, a hectare is approximately 40% of an acre. So Winnie the Pooh's Hundred-Acre Wood is about the same size as the Cavor School's 250-hectare campus.

"Hey! I think this is where I went to preschool!" said Dana. "Let's take a look inside."

"You do see where it says 'Keep Out' in big unfriendly letters, right?" said Noah.

Dana glanced around them, and, seeing no one in sight, tried the door handle. It clicked softly and the door opened a crack.

"Well, they really should have locked the door if they didn't want people to come inside," she said. "Come on." She slid through the doorway, and Noah followed her.

"You are going to get us expelled from this school before we even get started," said Noah. He usually registered his concerns when Dana did things like this, but it rarely was enough to stop either of them.

"Don't be such a 'fraidy cat," said Dana, looking around the one-room building, which was divided by child gates into a play area, a reading corner with several child-sized chairs and two bookshelves, and an office area with two adult-sized desks and a file cabinet. The light from the shaded windows was filtered and dim, but enough to see the walls decorated with letters, numbers, planets of the solar system, and other colorful educational posters. Dana tried hard to remember ever being here before.

She walked over to a bookshelf and picked up a book. *Peach Mountain Friends — The Science of the Seasons.* It was about a cat scientist, a monkey hacker, a warthog engineer, and a deer mathematician. She felt a flicker of memory. Monkey had been her favorite. And hadn't there been a cartoon too? Looking at it now, she could see it was more than merely entertainment for rowdy toddlers. Each character represented one of the STEM disciples: Science Cat, Hacker (Technology) Monkey, Engineer Warthog, and Math Deer.

Dana thumbed through the dog-eared pages, pausing at a verse where Cat ponders how Earth's axial tilt causes weather changes:

Cat thinks, when not sleeping,
Of Earth's axis keeping
A sharp tilt at its core:
Twenty-three and point four!
That's a lot of degrees!
It gets dark, and things freeze!

"No wonder I like math and science so much!" She laughed. "I was brainwashed early!"

"Ha," said Noah. "There are worse things." Noah was looking at the other bookshelf. "Look at the books over here!"

"*Mapping Your Memory Palace, Achieving Instant Meditation States, Distress Tolerance for Juveniles, Coping Skills Bingo*?" He read a few titles. Dana looked at the shelf and could see what looked like advanced textbooks on brain chemistry and neuroscience too.

"If they were teaching this stuff to pre-schoolers," Noah said, shaking his head, "that would seriously explain a lot about you. Can we get out of here now? I'm not too comfortable with this trespassing thing."

Dana ignored him and made her way over to the adult-office area. She opened the file cabinet drawer labeled "Students."

"Look here," she said, holding up a folder. "'Dana Dash.' I wonder who else went here." She reached for another folder labeled "Attendance" and pulled out a list of students. There were only ten names, hers included.

"Who cares?" said Noah. He was tugging at his shirt hem. "Come on, let's go before — " He stopped speaking abruptly, his attention suddenly fixed on something on the desk.

"What is it?" asked Dana.

"Shhh..." said Noah. "Be quiet and still."

Slowly and smoothly, he unslung his backpack and pulled out a small, clear container and opened the lid. "This time I'm ready for you," Noah said, moving carefully towards the desk. Dana saw what he was looking at. Another of the mystery glass lizards was perched on the edge of the desk.

Noah lunged for the lizard, but it was quicker than he was, and leapt off the desk, over his outstretched hand. Flaps of skin between its legs opened up and Noah's jaw dropped as it glided through the air. The lizard was flying!

As it went past Dana, her hand darted out and she snatched it out of the air. "Wow! Good moves!" said Noah. "Here." He held the open container towards Dana. She dropped the little lizard in and Noah sealed the top. He held it up to the light for them both to see. The brownish color it had been on the desk was already fading to glasslike transparency in the plastic.

"I've never seen one do that before," said Dana.

"I can't believe this," said Noah. "It has the same translucent skin and it's the same size as the one at your house. But I know that one wasn't a glider. This one is like a southeast Asian lizard called *Dracos Volans*, but again, smaller. And how can there be another species here doing the same transparent chameleon thing? Something seriously strange is going on here."

Noah stared at the captured lizard for several minutes, mumbling to himself about biological impossibilities. Dana looked at the little creature too until her head snapped up abruptly. She looked towards the door. "I think I hear someone outside!" she whispered.

Sure enough, the door handle was turning. Dana closed the file drawer and shoved the class list she was still holding into her pocket. She ducked behind the desk just as the door opened, pulling Noah's arm so he crouched down with her. Noah stuffed the sample container with the mysterious lizard into his backpack.

"Who is here?" The voice, a woman's, seemed to fill the small room. The speaker stepped into the daycare building and the floor groaned. There was a pause, then heavy steps moved slowly and quietly, like a hunter, towards the place where the two children were crouched. Dana motioned to Noah and they each crawled under a desk.

Dana peered out from her hiding place as a tremendous figure blocked the sunlight from the shuttered window. At first, the sheer size of the outline made Dana think this was a different person than the one who had spoken — a huge man — but, as her eyes adjusted, she realized she was looking at an enormous woman, almost two and a half meters tall.[28] She wore soft leather boots, denim overalls with more than the usual number of pockets, and a red plaid shirt rolled up to her elbows to expose meaty forearms. Her blonde hair was wound around her head in an intricate weave of braids. She also had a patch over one eye, which, combined with the braids and the overalls, made her look like a Viking-pirate who had migrated west to become a lumberjack.

Her huge hands — which looked like they could palm trees and oars and Dana's head like a basketball — had knuckle tattoos. Rather than the usual cliché LOVE and HATE, they seemed, incongruously, to be mathematical symbols. Each of this lady's large fists packed the punch of a powerful equation.

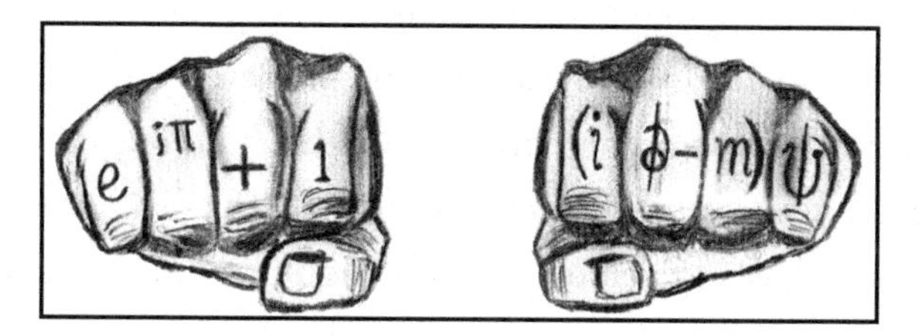

From the notebook of Dana Dash

Dana recognized one of the equations as Euler's identity[29] but the other wasn't immediately familiar to her. It had the Greek letter *psi* and something like the Leibniz notation[30] differential symbol, but with a slash through it. She did her best to commit it to memory to look it up later.

---

[28] Maybe eight feet tall —way taller than the tallest person you've ever seen.

[29] This famous Identity, written as $e^{i\pi} + 1 = 0$ or $e^{i\pi} = -1$, is considered one of the most beautiful mathematical equations, in which Leonhard Euler (1707-1783) revealed a simple and profound relationship between several important mathematical constants.

[30] Gotfried Willhelm Leibnitz (1646-1716) is credited as having developed calculus independently of Isaac Newton. He created a notation that is often thought to be more natural and useful than Newton's.

The immense woman walked past the children's hiding place. The ceiling was barely high enough for her to stand upright, but she ducked and dodged around the fluorescent tube lights. She was talking to herself. "Moreau saw children, he said. He is never wrong about what his little friends see. But I see them not. Through the back door they escaped, perhaps?"

She lumbered towards a door on the far side of the room, making the desks Dana and Noah crouched under tremble with every footstep.

"Run!" yelled Dana, scooting out from under the desk. Noah followed immediately behind her, and they sprinted for the front door.

"Ho!" boomed the woman, her voice echoing off the walls. "Halt!" She thundered after them, the whole building shaking with each heavy footfall. But two trespassers were already out the door and running at top speed across the campus lawn. From the doorway, the giantess yelled angrily after them. "Heed the sign and in the future, keep out!"

Dana and Noah didn't slow down enough to even glance behind them, and soon they were around the corner of another building and out of sight. Dana laughed as she tried to catch her breath. "That was a close one!"

"It's not funny," said Noah, between gulps of air. "We almost got caught. And have you ever seen a woman that big before? Imagine what she could have done to us if she had caught us!"

Dana laughed a little more at the thought of the big woman throwing each of them over an enormous shoulder and carrying them off like sacks of potatoes. "I've never seen *anyone* that big before! But I think we're ok. She only saw us from behind. I don't think she could pick us out of a lineup."

"You joke, but one of these days we probably will wind up in a police lineup. We could seriously go to jail, if we aren't more careful," said an annoyed Noah. "I told you we should get out of there right away."

"You certainly did," said Dana. "Then you spent a bunch of time looking at your new lizard friend instead of leaving."

"Oh, right, the lizard!" said Noah.

He started to unsling his backpack, but Dana stopped him. "Wait," she said. "We're still on a mission. Let's do what we came here to do, and you can look at your new pet all you want when we're done."

Noah nodded in agreement, and they continued across the Cavor School campus to the manor house with laser focus and no further detours. When they got there, Kevin greeted them much the same as before, and this time they were able to wave goodbye as enthusiastically as he did. As they left, they passed the quaint vegetable garden again. Dana noticed that the scarecrow had lost its hat, and looked a bit forlorn without it, like it didn't know who it was anymore. Its head hung limply to one side, and the painted-on eyes looked especially vacant in the bright afternoon sun.

Dana and Noah walked together along Big Oak Drive without their usual competitive energy, each lost in thoughts of the Cavor School and the contest. Eventually they said goodbye and Dana headed home alone, feeling anxious but hopeful.

CHAPTER 7 — THE ENVELOPE PLEASE

The final week of school seemed to take a whole year. Each morning, Dana would wake up and steel herself to face the day. Principal Peters and Matt McCaws had succeeded in convincing the entire school that Dana was, in fact, a dangerous presence who had to be officially tolerated, but should be, in every unofficial way, shunned. Except for Ms. Astrulabi, the teachers eyed her every movement with suspicion. And except for Noah, the students ignored her almost completely. She was even the last person picked for ball hockey — a clear sign that her popularity had sunk to the bottom of the school's social river. She had always been one of the first picked, and one of the best players, but nobody but Noah would pass her the ball now, and on the last day of gym class, when she scored the winning point, nobody cheered. She had never been more grateful for Noah's loyalty, and as far as she was concerned, the school year couldn't end fast enough.

The last day of school was the day of the science fair. Banned from entering but allowed to attend, she watched in silent frustration as Matt McCaws was awarded first prize for his project, which was a demonstration in making play slime. *His whole project was taken from an online video. He didn't even explain the chemical reaction of the ingredients — he just made something purple and sticky.* Dana had done this herself in her kitchen, and even those results had been better than Matt's prize-winning goo. When she got home that afternoon, she checked the mail hopefully for a letter from the Cavor School, like she had done every day that week. And like every day that week, there was nothing in the mailbox for her. For the first time in a long time, she felt herself fighting back tears. *What if....* She couldn't even finish the thought.

She thought about calling the school, in case the notification had gotten lost. She called Noah instead. He was much less anxious about it than she was.

"If we don't make the cut then we don't make the cut," he said. "Hillhurst Middle School will be fine. They'll still have to let you take science classes, even if they won't let you do any experiments or extracurricular things. They can't stop you from learning, right?"

Dana frowned. Of course *he* wasn't anxious. He wasn't the one being punished with a science ban till the end of time, or until graduation, whichever came first.

"Ugh, you sound like my parents. They're always telling me to relax and be more like a kid. That being a famous scientist can wait until college."

"Ha. Well, speaking of which, you know that lizard that I said was totally impossible and would be a great new discovery if I could document it?"

"The one that flew!" said Dana.

"Right. Well, it impossibly disappeared. Like the other one."

"I told you they were tricky that way," said Dana.

"Yeah, but this time there's no way it could have escaped. Zero. It was in a box with a heavy book on the lid."

"So you think it vanished into thin air?"

"Well," said Noah. "Actually yeah, sort of. You know what I think? This is going to sound a bit crazy, but I actually think it burned up."

"What?" said Dana. "Burned? That sounds more than a bit crazy. That's totally looney tunes."

"Hear me out," said Noah. "Sealed box, like I said. The next morning, no lizard. The box is clear plastic, and I could see into it through the side. In the bottom of the box there was ash and I swear it was in the shape of the lizard. The plastic underneath the body outline is discolored, blackened and lumpy, melted a bit."

"I'm hearing you," said Dana, "but it's a little hard to believe. How would that happen?"

"No idea," said Noah. "I have no scientific explanation for this. I'm a little freaked out. But I seriously need to catch another one of those things."

"Right! Repeat the experiment," said Dana. "The most important rule of scientific inquiry. No matter how weird the hypothesis." As she said that, something tickled the back of her mind. An experiment she wanted to repeat? Oh yeah — the umbrella — the stone. But she would need to be at the Cavor School to have any chance of touching that stone again.

"Right!" said Noah. "But I've never seen any of these glass lizards around my house. Just at your house, and then again at the Cavor School. Although maybe a different species."

"Well," said Dana, "tomorrow is Saturday. Come over to my place. We can hunt around in my backyard."

"I was hoping you would say that!" said an excited Noah.

When the mail arrived on Saturday, Noah and Dana were on hands and knees, searching through the bushes in her front yard. She didn't notice when the mail carrier rang the bell and had her father sign a form. It wasn't until the mail carrier called goodbye to Dana and winked at her that she looked up and saw her father holding the large blue envelope.

"Noah!" she cried out excitedly, and scrambled backwards out of the bushes onto the lawn.

"You found one?" Noah looked up from where he was foraging, his dark hair stuck with bits of leaves and twigs, possibly an insect or two.

"Better! Look!" She called back as she ran up to the porch. "Anything for me?" she asked.

"Hmmm," said Mr. Dash. "Let's see... Yes, I believe we have something for a Ms. Dana Dash and a Mr. Noah Knight." Dana's pulse quickened. "I'm sure it's nothing important," Mr. Dash continued, holding the envelope in the air just out of Dana's reach. "Do you want to open it after dinner? We could go get ice cream first?"

All she managed to say was, "Giveitgiveitgiveit!"

But her father was laughing and already opening it, saying "Let's see what we have here..."

"Hey! That's my mail!" said Dana.

"Ok, hold your horses. Hmm..." He was scanning the letter. Then he said, "Oh, Bug. I'm so sorry..."

Dana's heart sank into her stomach like a stone. There was a roaring in her ears and she barely heard him continue.

"... it looks like you won't be able to hang around with me all summer. You're going to be spending July at the Cavor School."

"You jerk!" she yelled at her father. Dana's father was smiling, and, she thought, she could see a twinkle of happiness in his eyes. "Why do you always do stuff like that?" But she was smiling too — then laughing. She grabbed the letter from her father.

"Really?" said Noah. "We made it to the finals? We get to build the biosphere?"

Dana read aloud from the letter:

Dear Ms. Dash and Mr. Knight,

Your proposal, "Self-Sustaining Bio-Sphere Lander," has been selected to compete in the Technology category of this year's Cavor School STEM contest. You are hereby invited to Phase II of the contest, wherein you will be given the opportunity to complete your project alongside other selected participants during the month of July. The final projects will be presented and judged in Phase III of the contest, and a scholarship to the Cavor School will be awarded to the winning team.

We congratulate you on your successful entry, and look forward to having you explore your ideas at the Cavor School this summer. Please have a parent or guardian call to make arrangements for your stay with us.

Most sincerely,
Dr. Catherine Cavor, Head of School

"Well," said Mr. Dash, "I think this calls for some celebratory pizza. As luck would have it, I have some in the oven. Come inside and eat."

July first dawned sunny and steamy, a perfectly perfect day as far as Dana was concerned. The first day of Phase II was to begin with a small ceremony at Myrtlegrove Manor with the participants, their families, some faculty, and a few other stray members of the community. As the Dash family pulled into the parking lot, Noah was waving at them excitedly from the side of the manor porch. He was wearing a smart red tie and a crisp white shirt, and his pokey dark hair was unusually slicked back, like his mom did for picture day. Dana looked down at her jeans and t-shirt and wondered if she was supposed to dress up.

Dana's dad got her small blue suitcase out of the trunk, and together the Dashes trundled to the entrance of the manor where Noah and his father, Mr. Knight, were waiting for them. Noah had three younger siblings, and Dana assumed that Mrs. Knight had stayed home with the little ones.

"Dana!" Noah was bouncing in his sneakers with excitement. "We're supposed to leave our stuff here." He pointed to a jumble of luggage next to the wheelchair ramp. "They'll take it up to the dorms for us so we can go straight to the science lab after the ceremony. We get to start right away!" Noah was grinning ear to ear and tugging at his belt loop, since his shirt was neatly tucked in.

Dana's father deposited her bag next to the others. The Dashes and the Knights entered the manor house's grand foyer, which had been set up as a small auditorium with rows of chairs facing the majestic staircase. A podium perched on the stair landing where the stairs diverged, above which, hanging on the wall, was a blue and white banner with the Cavor School's coat of arms. It had a symbol on it which Dana thought looked like the astrological symbol for Venus with the circle part containing the school motto, "DO IT WITH THY MIND." Above the crest, it said, "THE CAVOR SCHOOL" and below was the date the school was founded, "1876."

From the notebook of Dana Dash

The Knights and Dashes took their seats together towards the front. Mr. Knight was excitedly telling her parents about his upcoming trip, and while her mom listened politely, her dad was genuinely enthralled. Noah sat next to his father, tugging at his shirt now, which was very nearly untucked. Dana turned around in her chair to look at the small crowd starting to take their seats.

In the very back of the room was a man leaning against the wall, showing no inclination to sit with the other attendees. Dana had noticed him as they walked in; he seemed to be going from group to group, saying something in a low voice and then quickly moving on. He was dressed in an ill-fitting suit jacket and mismatched suit pants, and had a bowler hat pulled down low, nearly covering one eye. He was carrying a notebook, and had a pencil stuck behind his ear, which he frequently removed to scribble something on the notebook, and then tucked back in place. He reminded her of a man in New York City who had sidled up to her and her dad, asking if they wanted a Rolex watch or a Gucci handbag. He looked decidedly out of place at an event for a children's science contest.

She continued to scan the audience, and saw Ms. Astrulabi's emerald green headscarf towards the back. Dana waved excitedly. Ms. Astrulabi smiled brightly and mimed applause. There were a handful of kids about

Dana's age scattered throughout the seats, and she guessed those to be the other teams, sitting with their families. Other adults were taking their seats in the very front row, and Dana assumed these to be faculty — or at least involved with the school somehow, if they were sitting in the front. There was a chalky-skinned woman with a tight white bun sitting motionless beside a tall figure with a glossy, deeply brown head, clothed in flowing black robes. In the next seat, a tiny ruddy-cheeked woman in a bright green jacket threw back her head and laughed, making her rather large boof of springy crimson hair shake like a bobble toy. She was talking to — *Oh no!* Next to the leprechauny woman, at the end of the front row, was a large figure wearing denim overalls and an elaborate crown of sandy braids. It was the giant woman from the daycare! Dana slunk down in her seat, hoping the woman wouldn't see her, then realized that the woman was going to see her anyway when she went up to the podium. But still, Dana stayed low in her chair.

A shimmer out of the corner of her eye turned her attention away from the audience. Dr. Cavor had appeared at the top of the staircase, dressed for the occasion, which, for Dr. Cavor, meant that she looked as if she had walked under a rainbow and the entire color spectrum between green and purple had fallen on her.

Her jacket was an iridescent sky blue with turquoise buttons. Her pants were also iridescent-ish, but a deep midnight. There was a spectacular assortment of pockets in various shapes and sizes covering the exterior of both the pants and the jacket and, Dana suspected, the interior of them too. Dana even thought she could make out a few small bulges of fabric in the frothy lacey cuffs and collar of her cobalt blouse, and the somewhat uneven cuffs of her pants, because why not have pockets there too. Dr. Cavor's shoes were teal-ish, flat-ish, snug-ish, boot-ish, with no laces or zippers or any other way to see how they came on or off. She walked down the stairs and her shoes made absolutely no sound on the marble.

"Ah, the Dashes of Dash-on-Hudson," Dr. Cavor said, coming up to the row where Dana, her parents, and the Knights were sitting. "I was very pleased to see Dana's name on the list of project proposals this year."

Mrs. Dash smiled and Mr. Dash shook Dr. Cavor's hand warmly. "So, you do remember us. It's been quite a while, and Dana was only in your preschool a short time," Mr. Dash said.

Dr. Cavor replied, "Oh, Dana made quite an impression on me. Do you remember me from when you were little, Dana? You did not say much last month when you came to pick up the contest forms." Dr. Cavor had that intense, piercing look in her odd eyes again.

Dana felt her cheeks flush at the memory. It wasn't like her to be shy, or at a loss for words. She'd been so distracted by that silly dream that she'd just sat there! Much like she was just sitting here now, with everyone waiting for her to speak.

"I think... maybe I remember you," Dana lied.[31] If she had met this strange woman before a few weeks ago, she was sure she would have remembered, even if she had been very young. Dr. Cavor, to use Dr. Cavor's phrase, made an impression.

Dr. Cavor smiled, and Dana couldn't be sure, but she thought that one of the woman's strange eyes winked at her, before she turned to Noah and his father.

"You must be Mr. Knight. And you must be very proud of your son." She shook Mr. Knight's hand. "It is a pleasure to see you again, Noah," she said.

Mr. Dash re-inserted himself into the conversation. "Catherine, it's great to see you again. These days I'm a caver, and you're a Cavor, so we should have a lot to talk about!"

Dr. Cavor looked puzzled until Mrs. Dash explained the joke. "He means he's a caver, as in a spelunker. He explores caves as a hobby."

"Yeah," Donald Dash said, "It's the reason I know that the word 'spelunk' is onomatopoetic![32]"

---

[31] Which she had observed, from watching grownups in social situations like this, was expected.

[32] An *onomatopoetic* word is one that mimics the sound it names, for example, "drip," "chirp,"

Mr. Dash laughed so hard he almost fell out of his chair, but he was the only one. It was not clear if this was because the others didn't get the joke or if they were too horrified.[33]

Dana decided this was a good moment to return to socially polite conversation. "We are very excited to have been chosen for Phase II, Dr. Cavor."

Noah looked equally eager to salvage the conversation from the awkward mess their parents had made. "We think our project has genuine scientific merit. Thank you for giving us this opportunity, Dr. Cavor."

"Yes indeed, children." Dr. Cavor looked at her watch, which appeared to be a large red button attached to a wrist strap, and didn't appear to have watch hands. "If you will excuse me, I believe it is time to begin."

Dr. Cavor walked up the first set of stairs to the podium and faced the crowd. "Welcome, one and all, to the Cavor School STEM competition." The buzz of conversation quickly stopped. "This competition dates back to 1996, the year that the Cavor School first started accepting male students to study alongside the female students we have taught here since 1876.

"For over one hundred forty years, we have taken pride in our cutting-edge math and science curriculum. Our contest is designed to model the typical process by which knowledge is advanced in the real world. The first phase is the submission of a proposal, the way a scientist or mathematician might write a grant, or an engineer or technologist might seek private funding. In phase two, the finalists here today will be given working lab space, supplies, and equipment to finish designing and demonstrating their ideas.

"I am pleased to introduce the teams and their projects in the four STEM disciplines: Science, Technology, Engineering, and Mathematics." She took an envelope from a pocket, and with Oscar-worthy dramatic

---

or "buzz." So the joke here is that cave explorers are "spelunkers" because of the sound they make when they fall.

[33] Jokes that have to be explained in the footnotes aren't funny. Or are they? Dad jokes should always come with footnotes.

slowness, opened it, removed a piece of paper, set it on the podium, and smoothed the creases. At least three members of the audience coughed awkwardly during this process.

Dr. Cavor straightened her glasses, cleared her throat, and finally said, "In the Mathematics category, Tevi and Toma Torrez."

The audience applauded as two dark-haired pre-teens walked up the steps and joined Dr. Cavor on the landing. They were not identical twins, but were obviously related: the same height, the same tawny faces and wide brown eyes, the same long hair, and nearly the same clothes: T-shirts and jeans, just like Dana, who was relieved she wasn't the only one who hadn't dressed up.

The names Tevi and Toma sounded familiar. She frowned and opened her project notebook to retrieve the folded class list she had found[34] in the old Moonbeam Daycare building. She pulled it out and unfolded it.

"In the Science category, Lemma Lei and Prima Patel," continued Dr. Cavor. The audience applauded as the Science finalists walked towards the stairs, but Dana wasn't looking at them. She was looking at the class list. Sure enough, both Tevi and Toma Torrez were on the list. And so were Prima Patel and Lemma Lei. Also a Lambda Lei, certainly related.

"Our Engineering category finalist, Jayla Jones." A girl with coppery skin and curly black hair brushed back into two thick braids started forward from behind the rows of chairs. The applause began to falter as people realized that this finalist was in a motorized wheelchair, proceeding towards a staircase with no obvious form of wheelchair access.

"And in the Technology category, Dana Dash and Noah Knight." The applause was still sporadic as people watched the girl in the wheelchair moving slowly towards the stairs.

Dana barely heard her and Noah's names being called. She was noting that Jayla Jones was also on the daycare class list. What was going on? Here was a list of ten children who, six years ago, had been at the same pre-school

[34] And taken.

together. Out of those ten names, six of them were finalists in this year's contest — all the finalists, in fact, except Noah, were on the list. Dana was skeptical of coincidences, and by her reckoning, this was way more than a coincidence.

But there was no time to think about it now. As they reached the staircase, Noah pulled her to a stop before the first step. Jayla was just arriving and Dana hesitated for a moment, looking up at the four other finalists standing above her, next to Dr. Cavor at the podium. Then she motioned for the other finalists to join her, Noah, and now Jayla at the foot of the stairs.

Tevi, Toma, and Prima saw what Dana and Noah were doing and quickly moved to join the new lineup. The applause picked up to full volume. Lemma, however, pretended not to see what was happening until Dr. Cavor looked up from her podium list and motioned to her as well. Lemma rolled her eyes and walked sulkily down the stairs to join the other finalists.

Dr. Cavor detached the microphone and also stepped down to stand with the children.

"Each student will now introduce themselves and tell us something about their project." She handed the microphone to Toma, the first finalist in line.

Toma took it hesitantly and Tevi reached over so that they each had a hand on the mic. Toma took a deep breath and looked up at the ceiling while Tevi looked down at the floor. In a trembling voice, Toma said, "We're Toma and Tevi Torrez and," Toma took a deep breath and Tevi jumped in, "...we're working on a cryptography system based on curves in non-Euclidean geometric spaces and," Toma continued, "...we still need to flesh out the math but," Tevi finished "...we're sure it will work." They paused, then together said, "Thank you." They both sighed in relief as they passed the microphone to Lemma.

"I'm Lemma Lei. My assistant Prima Patel and I will be running experimental trials to prove my theory that animal behavior can be predictably influenced by external electromagnetic signals." Lemma paused after she finished speaking, as if waiting for a round of applause. When somebody in the front row coughed, Lemma frowned, and handed the microphone around Prima to Jayla.

Jayla said, "My name is Jayla Jones. My project involves a design for smaller, cheaper neurofeedback circuits, for use in prosthetics and other cybernetic systems, allowing direct brain control of machinery." She gave Dana a small smile as she handed her the microphone.

"Hi! I'm Dana Dash." She put the microphone under Noah's mouth.

"And I'm Noah Knight," he said.

She continued, "For our project, we intend to build a working environmental module for a mock spaceship."

She put the mic back in front of Noah, who said, "And we plan to test it ourselves by living inside it for a week. Recycling air, water, and wastes in a completely self-contained biosphere."

Dana moved the mic back to her and said, "A week is about the time needed for a lunar mission, so a week inside this biosphere will prove that our design could be used for real space travel."

Dr. Cavor crossed over to get the microphone. "Four excellent projects. Four excellent teams." She paused for applause, and this time, the audience responded to the cue enthusiastically. "Thank you all for coming. I invite you all back next month to see their proposals come to life. Now, finalists and their families, please join me outside for a tour of the campus."

Dr. Cavor walked purposely towards the door, and the finalists trailed behind her like a row of ducklings. The audience clapped politely and remained seated; this procession was part of the ceremony, but Dana's parents, Noah's dad, and the other finalists' families looked a little unsure of what their role was. A woman seated in the front near the faculty was the first to rise from her seat, followed by a girl who looked nearly identical to Lemma Lei but a little taller. *Could this be the Lambda Lei from the list?* Dana thought. Dana, last in the row of Dr. Cavor's ducklings, glanced behind her as she walked out the door, and was a little relieved to see that her parents and the other families were following the lead of the woman who was probably Lemma's mother, and the girl who was certainly Lemma's sister, although they were proceeding towards the door in a much less orderly fashion than the children had done.

She also saw Ms. Astrulabi motioning to her. Dana paused in the doorway and waved back. Ms. Astrulabi shook her head and, glancing furtively around her, got up from her seat in the back row and hurried towards Dana. She ushered Dana onto the manor porch, ahead of the gaggle of parents migrating towards the doorway behind Lemma's mom.

"Dana, I wanted to give you something. A gift from... a gift for you." She thrust a small bronze disk on a chain into Dana's hand.

"Oh wow, Ms. Astrulabi. Thank you," said Dana, eyeing the object curiously. It was unlike anything she'd ever seen before. It had an inner wheel and slider, and was covered in strange curves and markings. It felt warm in the palm of her hand.

"It's an astrolabe," Ms. Astrulabi said in a low voice. "It's used like a sextant to sight stars, planets, and the sun, for navigation or to tell time. A thousand years ago, this was like having a smart phone. My name is actually an old professional name like Smith or Baker; it means someone who makes devices like this.[35] I thought you should have one for — for whatever adventures await you in the future."

"I'm... That's so... thank you, Ms. Astrulabi. I'm going to miss you next year." Dana put the chain around her neck. It was surprisingly lightweight. She was eager to play around with the device, but right now —

"Dana?" She turned to see Noah waving at her from the flock of finalists and families waiting with Dr. Cavor, who had started leading the group down a brick pathway. Her parents were looking around for her. Dana turned back to her favorite teacher, to say goodbye and thank her again. But Ms. Astrulabi had already gone, lost into the small crowd now

[35] Mariam al-Alijliyah "Al-Astrulabi" was a famous astronomer who lived in Aleppo a thousand years ago. She was renowned for her astrolabes, a tool for navigating by the stars.

meandering out of the manor house.

Dana tucked the astrolabe under her t-shirt and hurried to join the others. She'd have time to play around with the ancient invention later. Right now, there was a modern science lab to explore.

CHAPTER 8 — THE SHAPE OF THINGS

Dr. Cavor led the finalists and their parents from the manor house down a pathway with wide swaths of well-manicured lawn on either side. Mrs. Dash was walking next to Dr. Cavor, asking questions about the history of Myrtlegrove Manor, while Dana and Noah's dads were having a very animated conversation about caving. The other adults walked in a cluster making adult small talk. The children followed behind, not talking. Dana, Noah, and Jayla were at the very back of the pack.

As they walked, Dr. Cavor pointed out features of the campus like a tour guide. There was the auditorium, built in 1901 as a small theater for the town. There was the greenhouse, which pioneered hydroponic grow systems to help feed needy people during the Great Depression. There was the Zen garden, the cafeteria, and the art studio. To the east were the indoor pool, gymnasium, and track field. Dr. Cavor gestured towards a long white hangar-style building with six garage-sized doors that each had a matching people-sized door.

"This building was added to the campus in 1995 with a large private donation that included the stipulation that the school open its doors to male students," she said. "Each set of doors you see here leads to an individual laboratory work space. Each team will be assigned a lab in which they can develop their project." She gestured again to another building up the hill from the labs. "Each team member will also be assigned a room in the Lower School dormitories where they can stay for the duration of the competition."

The gray-haired woman who had accompanied Jayla Jones asked, "There's more than one school here?"

Dr. Cavor explained, "The Lower School is the school for the lower grades, five through eight. The Upper School is grades nine through twelve.

These are usually called middle and high school in the solar system... er sorry, the public system. We also have special after-school classes for exceptional students of all grades, held at the manor house, and since those classes are held after dinner, we call it the Moon School."

Dana repeated "the Moon School" under her breath and locked eyes with Noah. "We have to win this thing," she said softly, "Then for sure we get into those special classes."

Noah replied, "We need to win a scholarship just to get in at all. Neither of our parents can afford tuition for this place. We'll be staying in public school if we don't win."

Jayla, riding nearby in her motorized wheelchair, overheard them. "Oh yeah, me too," she said. "I live with my grandmother in the Bronx. If I don't get a scholarship here, it's Bronx Tech for me. It's not bad, especially for a New York City public school, but it's nothing compared to this. This place is something special." She looked over the campus hills and gardens leading towards the lab building, which gleamed so white in the midday sun that it almost seemed to glow. "You two are lucky you have each other to team up. My only science-y friend moved to London."

"Hear that, Dana? You're lucky to have me," Noah said, and then to Jayla, "It's nice to have a partner with a lot of energy, but she can really overdo it sometimes."

"Ha!" said Dana. "Please tell me all about how I overdid it after we win!"

"Well, good luck to you both," said Jayla.

"You'll need the luck, not us." Dana said. She regretted saying it immediately.

"Okaaay then..." said Jayla, a little taken aback.

"I didn't mean..." Dana trailed off. Jayla said nothing.

Dr. Cavor had been fielding questions while the three children had been speaking in the back of the group, and the children heard none of the answers. Now she was saying, "...inside the science labs," and the group moved forward again.

Dr. Cavor led them through the second of the human-sized doors, labeled "SL4." She punched a code into the keypad. When the door clicked open, the whole group scrunched into the room, set up with four desks and computers. It was not small, but with the seven children and ten adults, the room was uncomfortably cozy for a July afternoon.

"This is the lab's office space," Dr. Cavor said, "and you can see the experimental lab space through the window there." She pointed. "The lab can be entered through the office here, or through the roll-up hangar doors. Each lab also has a large retractable roof section, which is why we sometimes call them the 'Sky Labs.' Some of you may understand the reference to the name of the U.S. space station that orbited Earth from 1973 until its re-entry and burn-up in the atmosphere in 1979."

She opened the door and they all followed her into the cavernous lab space. Along the back wall was a row of metal lockers, and eight gleaming white lab tables were set up, spaced evenly across the room. Four were set up like work desks, with computers and chairs, and the other four like lab stations with stools.

Dr. Cavor continued, "The labs assigned to each team are identical. Each has a collection of standard equipment, but students may request any additional supplies as needed. Lab Five, next door, is my personal workspace, and Lab Zero belongs to Professor Moses Moreau, Master of Sciences here at the Cavor School."

"Lab Zero?" asked the man standing with Prima; Dana assumed he was her father.

"We start our addressing systems from zero rather than one here," said Dr. Cavor. "Lab Zero is not special. Or, at least no more special than my own Lab Five. Take note, children: those labs are entirely off limits when we are not here. When Professor Moreau or I are working in our labs, we do not wish to be disturbed, so please do have a very good reason if you feel the need to interrupt. Call first." She gestured towards a dusty desk phone, which had a dog-eared campus directory next to it.

"This is Science Lab Four, which I am assigning to Ms. Dash and Mr. Knight. You and your parents may stay here and look around while I show the other teams their lab spaces. I will return to answer any questions you might have. The keypad code for your lab's door is written down on a note in the drawer of the office desk closest to the door."

Dr. Cavor left with the rest of the group in tow. Dana's parents and Noah's father ambled around the lab, opening drawers and cabinets, noting the bathroom and state-of-the-art safety equipment. Then they wandered back into the office, where there was much less to look at. Their parents tried to cajole them into exploring the grounds, but Dana and Noah were itching to get started in the lab. Who wanted to sight-see when there was a spaceship to build?

So their parents hugged them goodbye, looking both sad and excited to have them out of the house for a month. Dana and Noah started methodically rifling through the cabinets and making lists of things they were going to need for their project. When Dana found a tape measure, she got out her project journal and began measuring walls and doors. After a few minutes, Noah came over to see what she was doing.

"Planning on moving some walls around?" he joked.

"It's just something I like to do," said Dana. "Anywhere I spend a lot of time, I like to map it on graph paper. I feel more comfortable in a space once I've mapped it out."

"Well," said Noah, "There are definitely messier ways to mark your territory."

Dana laughed and said, "It started with a treasure-hunting game my dad liked to play with me. He does it every now and then, but he came up with the idea for our yearly Easter egg hunt.

"We drew our house plan out on three large sheets of graph paper, one for each floor. Then we assigned a coordinate system to the graph. The X axis went approximately west to east, and the Y axis south to north. The Z axis was down to up. Each square on the graph was one quarter of a meter.[36] So any spot in the house could then be located with three numbers — X, Y, Z, coordinates. My dad put the origin in the kitchen, right at the handle of the refrigerator door. He said that he knew that was the origin, because he always wandered back there when he wasn't doing anything else." Noah grinned appreciatively.

"When I was little he just used basic arithmetic problems for the clues, but as I've gotten better at math, I get equations for curves and I have to find the intersection points."

Noah said, "Wow! That is a really imaginative way to teach your kid math. Your dad has some seriously corny dad jokes, but your parents are really cool."

"Yeah," said Dana. "They're ok, I guess." She thought about how last year her parents, for the fourth of July, staged a historical treasure hunt for her around Dash-on-Hudson, using her mom's love of American history and her dad's love of maps and games. She would be here this year, so they wouldn't get to do it again. She felt a twinge of sadness at the thought but shrugged it away. "Anyway, I think my Dad was just mad that Easter moves around, making it impossible to pretend he is celebrating some famous scientist's birthday. So he came up with a new math-based tradition instead."

As Dana talked about her rebelliously scientific family traditions, Noah was looking at the drawing she had made of the lab and office. He tugged on his shirt tail (which had long since come untucked), and then turned abruptly and walked to the back of the lab towards the bathroom adjacent to the office. He peered behind one of the shelving units.

"Hey!" Noah said from the corner of the room. "Come give me a hand!"

"What?" Dana glanced up from her notebook to find that Noah was pulling books off the shelf. "What are you doing?"

"I think there's something behind here!" He was now trying to move the shelf.

---

[36] About 10 inches.

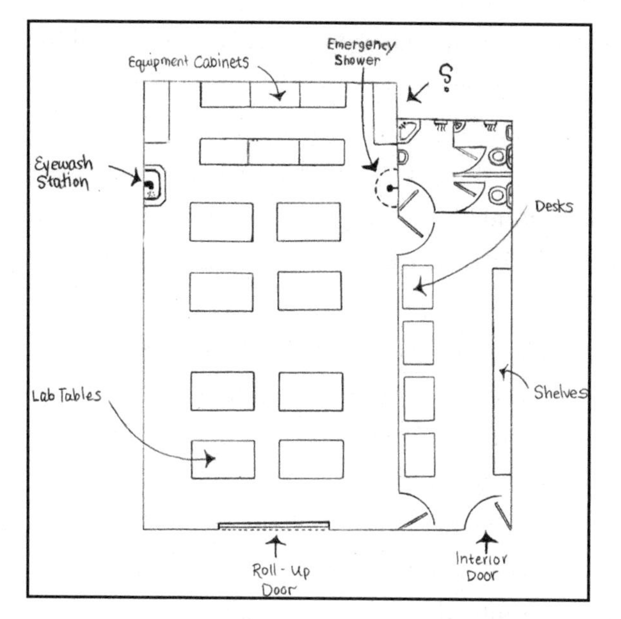

From the notebook of Dana Dash

Dana, obviously supportive of shelf-moving if there were mysteries involved, joined in. They managed to shove the shelf away from the wall a quarter meter. Just as Noah had suggested, there was something behind the shelf: a door.

"How'd you know?" Dana asked.

"Look at your map. The bathroom next to the office doesn't extend as far back as the rest of the lab does. I figured that something else must be here."

"Oh," Dana said, "Certainly a storage area..." Dana tried the handle and the door opened noiselessly, revealing a dark hallway. "...or a hallway," she concluded.

"It must connect to the lab next door. That's Dr. Cavor's lab!" exclaimed Noah. They stood in silence for a moment, staring into the dark passageway.

"Does she know about it?" asked Noah.

"I'm sure she knows every centimeter of this school."

"So why wouldn't she lock it?"

This was a very good question. Dana, personally, didn't like *anyone* rummaging through her things.

"I... I'm not sure."

"Wow, I don't think I've ever heard Dana The Great admit that she wasn't sure about something!" Noah mocked, his smile lopsided. "Does it hurt to be in the dark? How are you holding up? Is the unknown killing you?"

"Shut up," Dana ordered crossly. "This is serious."

Noah rolled his eyes at her. "Yeah, I know. I was just joking. I say we scoot these shelves back and pretend we never saw it. Dr. Cavor specifically said *not* to look through her office."

Dana nodded, but her curiosity about what was on the other side was burning. Dr. Cavor was so mysterious — who knew what she had in her lab? Maybe that umbrella with the strange stone. She remembered the tattered figure from her vision reaching for her and backed up fast, causing Noah to stumble, and quickly pushed the shelving unit back in place.

"That... worked on you?" Noah said hesitantly.

"We should focus on our project. No time for so-called secret passageways that probably just lead to broom closets." But she whispered to herself, "At least not right now."

After surveying the lab, Dana and Noah decided that the office was the best place to do their project planning work. They sat across from each other at one of the desks, and Dana opened her notebook to a clean page.

"First order of business," said Dana, "is how to build the frame for our new biosphere. When we sketched it out for our proposal, it was a sphere on legs, but I'm not finding any giant industrial-strength hamster balls for purchase online."

"We named it the Bio-Sphere Lander," Noah said. "If it's not going to be a sphere, we'd have to change the name, and 'Bio-Cube-Lander' doesn't have the same ring to it."

"Well, we can get more sphere-ish, and still build it with regular-sized pieces. Look." She drew five shapes: a pyramid, a cube, a double pyramid, and then two more shapes which looked like RPG dice, with twelve and twenty sides respectively. "These are called the Platonic solids.[37] They would each give us an enclosure made of same-shaped faces, and same angles, so construction would be fairly simple." She pointed to each shape.

"Tetrahedron, hexahedron, octahedron, dodecahedron, and..."

Noah interrupted, "...icosahedron, I know."

Dana was annoyed at being interrupted. Certainly more annoyed than Noah was at having something explained that he already knew. She shook it off, reminding herself that she must have done that to him a hundred times. He sometimes teased her for "Danasplaining" but he never lost patience with her. She quickly jotted down a table showing the number of faces, edges, and corners for each shape. She also included columns for how many edges met to form each corner and the number of sides for each of the polyhedron's faces.

PLATONIC SOLID	REGULAR POLYHEDRON	FACES	EDGES	VERTICES	VERTEX SIDES	FACE SIDES
	TETRAHEDRON	4	6	4	3	3 TRIANGLE
	HEXAHEDRON (CUBE)	6	12	8	3	4 SQUARE
	OCTOHEDRON	8	12	6	4	3 TRIANGLE
	DODECAHEDRON	12	30	20	3	5 PENTAGON
	ICOSOHEDRON	20	30	12	5	3 TRIANGLE

From the notebook of Dana Dash

She said, "Do you see it? When you look at the numbers lined up like this, there are some interesting relationships between the shapes."

[37] Plato discovered these before much science was really invented, so pretty impressive.

"Yeah, I can see it now," said Noah, "The cube and octahedron numbers are kind of cousins. And the dodecahedron and icosahedron too."

"Right," said Dana, "Looks like if you cut the corners off one shape, in just the right way, it turns into its related shape. And the tetrahedron numbers are symmetrical. It's its own cousin. You can cut corners off it to form another tetrahedron."

They both sat and looked at the table for a while, quietly admiring the elegance of the symmetries.

Finally, Dana said, "Also, I'm thinking that for strength, if we're going to pressurize this thing, we should run rods through the vertices. Your magic-algae-oxygen-farm will be in the center. So we need to think about the space we'll actually have left to crawl around in for a week, and how big to make it for minimal comfort."

"Hmmm," said Noah, "The icosahedron is the most sphere-like, but I can't imagine it would have a lot of space with all those rods through it."

"Actually, the dodecahedron would have more rods through it. It has more vertices, and the rods will make it stronger. But the icosahedron is closer to spherical, so the internal pressure will be more evenly distributed. Either one could be good. And we'll want something tough for the windows."

"Well," said Noah, smiling, "This big window between the office and the lab looks seriously tough. Probably so you can monitor dangerous experiments from in here."

Dana leaned over the desk and thumped the window. "Actually, this window material might be perfect for our windows. I'll ask Dr. Cavor about getting some."

"I'm all for the strongest choice," said Noah, "I remember your demonstration about how air pressure is no joke."

"Dodecahedron or icosahedron then. Let's build some models and see what we like."

After playing around with toothpicks and styrofoam balls meant for modeling molecular compounds,[38] they discovered that the office computers had some excellent CAD[39] software. The children electronically sketched several potential versions of their project, finally deciding on the twelve-sided structure. A smaller dodecahedron in the center would house the tanks for Noah's cyanobacteria[40] with support rods connecting each corner of the baby dodecahedron to the mothership. The whole thing would be mounted on five legs connected to a pentagon-shaped base plate, with enough clearance that they could crawl into the biosphere from an airlock hatch at the bottom. They agreed that this looked futuristic and super cool, very much how a spaceship landing module should look.

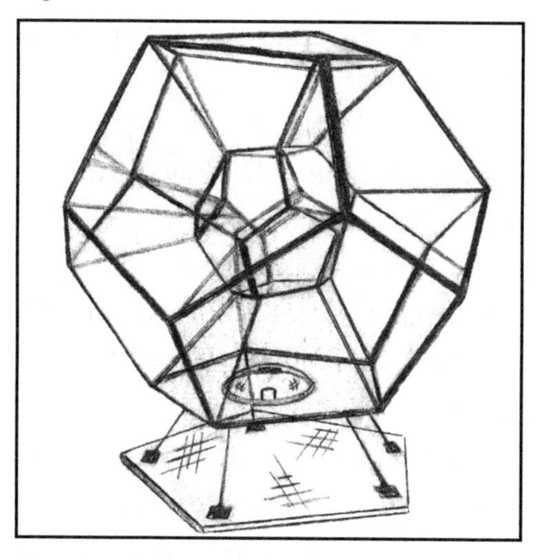

From the notebook of Dana Dash

They admired their progress, and then Dana's stomach growled. They had completely lost track of time. "That's probably good for today," she said. "I wonder if we should wait here for Dr. Cavor? She said she

---

[38] For example, water is a compound made of one hydrogen molecule (H) and two oxygen molecules (O) to form H_2O. If you make it with the model Noah and Dana have, it looks like this: o—O—o

[39] *CAD:* Computer-aided design

[40] Or as Dana calls it, magic-algae-oxygen-farm.

was coming back, and I want to give her our list of supplies as soon as possible."

Almost as if waiting for her cue, Dr. Cavor appeared at the door.

"Hello, children. It looks like you have had a busy day," she said, eyeing the abandoned styrofoam molecule models and discarded printouts of rejected biosphere designs. "Is there anything I can assist you with at this time?"

Dana, somewhat startled at Dr. Cavor's sudden appearance, handed her their list of supplies. The Head of School studied it thoughtfully.

"Two scuba tanks and a regulator?" Dr. Cavor asked.

"We need a way to leave the biosphere, in case we need to do repairs or something. Like a space suit. So we can't cheat. Extra valves are to maybe rig up a pump for the airlock."

"Yes, indeed," Dr. Cavor nodded, continuing down the list. "Thick plastic like the lab windows? Hmm." Dr. Cavor glanced up at the window in question. "That is transparent polycarbonate. I believe when we upgraded to a thicker version a few years ago after... an incident, we kept some sheets of the old thinner material. I think it is stacked in my lab somewhere."

"Great," Dana said. "The sooner the better. I want to test it."

"Angled aluminum[41] bar stock as close to 116.56 degrees as possible." Dr. Cavor pondered it. "That is certainly a specific request."

"That's the angle between dodecahedron faces. Our biosphere will be dodecahedron-shaped, with legs like a lunar lander. The bars are flat pieces of metal folded lengthwise at that angle to hold the windows in place, basically acting like both the frame and the support brackets, but stronger when it's one single piece of metal."

"I see. I will arrange for you to have all the items on your list with all due haste. I do look forward to seeing this Bio-Dodecahedron Lander

---

[41] The American word is "alumin*um*," whereas the British say "alumin*ium*." Language evolution like this can be very interesting — but one is not more correct than the other. And for the record, while the British usage is consistent with the neighboring elements sod*ium* and magnes*ium*, the American word came first.

completed." She folded the list and it disappeared into one of her many suit pockets. "*Bio-Dodecahedron Lander* — that is certainly not a name that rolls easily off the tongue."

"Hmmm..." Dana pondered this. "I guess we could use an acronym. We could call it the B.D.L.?"

Noah chimed in. "If you say BDL fast it sounds sort of like 'beetle.'"

"Ooh, yeah!" said Dana. "The Beetle!"

"I love beetles too," said Noah. "There are hundreds of thousands of different species of beetles."

"Yes," Dr. Cavor laughed. "One of the BDL's creators does seem to have an inordinate fondness for beetles.[42] So, the 'Beetle' it is. As your advisor, I will periodically stop by to see how you are progressing. Free to call me if there is anything you need. Or..." Dr. Cavor paused ever so slightly, "...if you just want to talk. But right now, children, you should be at dinner. My apologies; I should have sent you to the Lower School cafeteria twenty minutes ago."

Dana grabbed a map of the campus from the desk as Dr. Cavor shooed them out the door. She pointed them towards a building at the top of the hill, and then turned and headed for her own lab door. Noah and Dana took off that direction, moving fast, trying to make up time. Of course it turned into another race.

---

[42] This is a quote from the famous biologist J.B.S. Haldane (1860-1936), who is said to have been asked by a priest if he couldn't infer something about the mind of God from his extensive knowledge of biology. He is said to have replied, "Any creator must have had an inordinate fondness for beetles," based on the fact that the beetle family made up more than half of all known insects, and insects represented nearly 80% of all known species at the time.

CHAPTER 9 — SIGMA, PHI, PI, AND DELTA

Dana burst through the cafeteria doors just ahead of Noah. ""I win again! Dash by name..."

She stopped mid-sentence. They were last to dinner; all the other finalists were already seated with their food. Everyone had looked up at their abrupt entrance. Lemma and Prima were sitting together at a long rectangular table in the middle of the room, while Toma and Tevi were sitting with Jayla at a round table close to one wall.

Dana was about to sit at a separate table, but Noah walked towards the math twins and the engineer, and Dana, as she often did in social situations, followed his lead.

"Ah, good. The last two students are here." A woman entered the cafeteria carrying two trays, each piled high with an assortment of food.

Dana recognized the woman from the welcome ceremony earlier. In fact, she was certain she had seen this woman a few times around town. She was very noticeable, and not easily mistaken for someone else.

Her skin and long hair were nearly the same shade of white, and her thin, sharply boned features made her look like some sort of ice queen chiseled out of actual ice. The expression on her face conveyed much the same warmth. She was leaning into the theme,[43] wearing a bright white pantsuit that almost made her look like she had some color in her skin by comparison. Almost. The only real pigment Dana saw was in the woman's eyes: a reddish hue that almost seem to glow, surrounded by all that whiteness. Even her eyebrows and eyelashes were snow-colored.

Noah spoke to himself under his breath, but Dana could pick out the words. "Completely amelanistic oculocutaneous albinism — a mutation

[43] Is that what Sheryl Sandberg meant when she told women to "*Lean In*"?

of the tyrosinase gene on chromosome eleven causing a total failure in melanin production..."

He might have had more to say to himself on the subject but the albino woman spoke again as she approached the table.

"Here," she said, setting the plates on the table in front of Dana and Noah. "You missed the orientation information about the dorms and cafeteria. I am sure you can work it out. I am Dr. Gretta Griffin. I teach organic, inorganic, physical, and bio chemistry here at the Cavor School. Currently I am being forced to play mother hen for the duration of the STEM contest. You are the Technologist team, yes? I trust you met with the Master of Delta, Dr. Cavor."

"Right," said Dana. "We talked to her, but what is Delta — "

"Very good," said the albino woman, and abruptly walked out of the room.

"That was strange," said Noah.

"Delta?" asked Dana. She addressed the other kids at the table. "What is that about?"

"She explained it earlier," Tevi started. Toma continued, "Apparently students at this school are divided into four Disciplines, based on the same four categories as the STEM contest. Each Discipline has a Greek letter assigned." Tevi pointed to the far wall of the cafeteria where four large banners were hung. Each had white writing on a different color background, with a symbol, name, STEM designation, and motto.

Jayla was eating her food. But she stopped cutting the meatloaf, wiped her mouth, and spoke. "There's a staff member who is Master of each Discipline, and is also the advisor to the corresponding contest team. Mine is Dr. Tempora Taxidiótis, Master of Engineering. I talked to her earlier today. She really knows her stuff."

"And our math advisor, Master of Mathematics, met with us," said Toma, followed up by Tevi saying, "An amazing mathematician named Udo Uachen. He's a monk too, which is cool."

"Who is your science advisor?" Noah called across the rows of tables to the science kids.

"Professor Moses Moreau," Prima called back from their table. "Master of Science. He knows everything about biology."

"Oh!" said Noah, his interest piqued. "I'd love to meet him. Sounds like you all have great advisors."

COLORS	SYMBOL	NAME	STEM	MOTTO
BLACK + WHITE	Σ	SIGMA	MATH	"Define & Derive"
RED + WHITE	Φ	PHI	SCIENCE	"Discover & Deduce"
GREEN + WHITE	Π	Pi	ENGINEERING	"Design & Develope"
BLUE + WHITE	Δ	DELTA	TECHNOLOGY	"Deploy & Disrupt"

From the notebook of Dana Dash

Lemma spoke up. "Your advisor is head of the whole school. That hardly seems fair. My sister says the tech discipline only plays with other people's ideas and doesn't really discover or create anything new. So you clearly need the extra help. How does the team she coaches not have an unfair advantage?"

"Good point," Dana quipped. "Why don't you go right now and talk to her about how biased and unfair she is."

Lemma scowled and Dana scowled back. It was the opposite of love at first sight.

"Hey," Noah asked the science kids, "Do you two want to join us over here? It'll be easier for us all to talk if we aren't yelling across the room."

"We're sitting in the Phi seats," said Lemma. "This is our section. My sister Lambda goes here and showed me when I visited last year."

Dana looked up at the banners on the wall and saw that rows of long tables extended from beneath each banner across the cafeteria. Lemma and Prima were, indeed, sitting in the row with the Phi banner at the head. The rest of the kids were at one of the round tables on the far end of the room from the banners, and didn't seem to line up with any of the rows.

"Well, doesn't look like we're sitting in any of the Discipline-specific sections," said Noah. "Maybe these are the staff tables? Anyway, this appears to be neutral territory. Why don't you come over?"

Prima looked at Lemma, as if to silently ask permission. Lemma thought for a moment, then nodded. "Ok. That seems fair," she said.

The science kids got up and brought their trays over to join the others at the round table. Prima caught her foot on a table leg and nearly dumped her tray of food into Noah's lap. He smiled and helped her steady herself, and she smiled shyly back and took the seat next to him. Lemma was left to sit next to the math twins. Dana was seated next to Jayla, as far away from Lemma as was possible at a round table.

"So, tell me about your project," Noah said to Prima. "What are you working on?"

Lemma cut in before Prima could answer. "I don't think we should be discussing our work with the competition," she said.

"Oh come on," said Prima. "It's not like we're really competing. We're not working on the same things, and at this point, all our projects will either win or lose on their own merits."

Lemma scowled, which Dana was starting to think was the permanent condition of her face, and said, "Only one team will be declared the winner and get the trophy. I intend for that to be me."

"Well, I guess there is a 'me' in 'team,'" Dana muttered.

"Ok," said Noah. "I'll tell you about my part of our team's project. I'm doing some genetic modification on cyanobacteria. The genes I'm updating have analogs in the chloroplasts of all green plants."

"That's cool," said Prima. "Do you think that plant cells partnered with a form of cyanobacteria? The same way our cells probably took in mitochondria,[44] which were previously a separate organism."

---

[44] *Mitochondria* are the powerhouses of human (and all other animal) cells. They have their own DNA, which means that there were once two entirely separate types of single-celled creatures that teamed up to form more complex (Eukaryotic) cells which now form multi-cellular animals (like human beings).

"That's probably true," said Lemma, finding something interesting about the other students at last. "Both types of organelles have a double membrane and their own separate DNA."

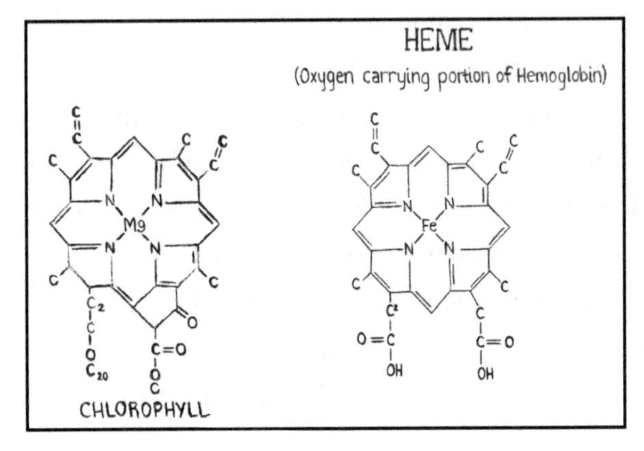

From the notebook of Noah Knight

"I think it's unlikely that this symbiosis occurred twice," said Noah. "Plants and animals probably have a common ancestor. You can also see that the similarities in molecular structures of significant molecules in plant and animal biology. Take a look at chlorophyll[45] and hemoglobin.[46] They're almost the same molecule. "He pulled out his lab book and sketched the two molecules for comparison.

Dana was having trouble following the plant science stuff. It sounded interesting, but bioscience didn't excite her the way astrophysics, electronics, and machinery did. Noah, Prima, and Lemma continued to talk plant DNA, but Dana's attention drifted. The math kids seemed to have tuned out the others, and were talking to each other in low voices. From what Dana could hear, it sounded less like language and almost like code. Or maybe very high level math. Dana thought she was pretty good at math, but their conversation didn't make any sense to her. She turned to Jayla, whose cybernetics project was much more Dana's sort of thing.

---

[45] *Chlorophyll* is the chemical in green plants that captures energy from light.

[46] *Hemoglobin* is the chemical in red blood cells that transports oxygen.

Then she remembered that when they had spoken earlier, she had been rude, albeit inadvertently. Dana chewed her food for a little while longer, working up the nerve to apologize in about the time it took to eat her broccoli.

"Jayla," said Dana, "About earlier, I'm sorry, I didn't mean to..."

Jayla stopped eating. "Hey. Don't worry about it. It surprised me, but I kinda liked it."

"What?" Dana was confused. "Why?"

"Being in a wheelchair means people are overly nice and sympathetic. It's a lot, you know? People always treating you with kid gloves, like you're super fragile. But your first reaction wasn't 'Oh this poor crippled thing.' Your first reaction was to see me as a competitor in a contest you want to win. To me that's a kind of respect. And I definitely prefer respect to pity."

"Oh," said Dana. "Ok. Well, I do intend to win this contest. But I hope that doesn't mean that we can't be friendly. Maybe even be friends."

"I'm willing to give it a try. Let's start over," said Jayla. She stuck out her hand and said, "Pleased to meet you, prospective new friend Dana."

Dana grinned and took her hand. "Pleased to meet you too, prospective new friend Jayla. So tell me, how's your lab? Did you get started on your project?"

"I mostly worked on setting up my radio today."

"Radio?"

"Yeah. I'm a ham. A licensed amateur radio operator. I talk to my friend in London, the one who moved. It's fun trying to get connections on various frequencies, using different techniques. Repeater satellites. Atmosphere skipping. Stuff like that. Right now we're working on setting up an EME. That's an Earth-Moon-Earth communication. If we get it to work, we can talk to each other by bouncing a signal off the Moon. The power levels are low but, with computer enhancement of the signal, we think we will be able to speak to each other."

"Wow!" said Dana, impressed and fascinated by Jayla's hobby. "That's wild. I've never played with radio stuff. Maybe you can show me sometime."

"Sure! If we can find the time between working on our projects I'll be happy to. Plus I can always use the help of someone with working legs. Maybe you can help me set up the EME antenna."

"I'm in," said Dana. "We can certainly make time at some point. It sounds like a great side project."

While they finished their dinner, they talked about how an EME antenna functioned and what they would need to build one.

"Shouldn't be too hard," said Jayla. "The lab is great. All the tools and electrical components a girl could ask for. Speaking of that, is it weird to you how much freedom we have at this school? Not that I'm complaining. I like being more independent like an adult. But our labs have welding gear and dangerous chemicals, and the adults don't seem to think anything of letting kids play around with it. Unsupervised. I know we're all young geniuses, but we're still only ten years old."

"Hmmm," said Dana. She looked around the cafeteria. "I use welding gear at home all the time, but my parents know I know how to do it safely. I guess most kids don't. And now that you mention it, we're here alone with no one watching us again. That's strange, especially for a school." Dana's eyes caught a small movement nearby and she said, "Well, almost no one..."

She leapt out of her seat, startling Jayla and everyone else as she lunged for a nearby table. "Gotcha!" She returned to her seat with her closed hand held out in front of her. "You got one of those sample cases, Noah? I got another lizard for you."

Noah pulled a small plastic case out of his backpack and Dana dropped the lizard into it.

"Crazy," said Noah. "This one looks like a tiny translucent iguana."

"Ah, those clear lizards," said Prima. "I've seen them around my house, but I've never caught one. They're super fast. Does it really look like an iguana? Lemme see."

Lemma said, "I've seen them too. Never identified the species."

“Has *everyone* seen these things before but me?” Noah asked.

They all said they thought they had seen the little glass lizards around too and that they must be indeed native to the area. The math kids, who hadn’t said much to the rest of the group, nodded in agreement in unison.

“I remember them from my old house,” Jayla said, “when I lived in Dash-on-Hudson before the... before I went to live with my grandma. I caught a few of them. They have to be native to the Dash-on-Hudson area.”

Noah started to protest, “There’s no record — ” but he stopped short as the albino woman came back into the room.

“If everyone is done eating, follow me to your sleeping quarters for the next four weeks. And clean up after yourselves. The composting and recycling are over there.”

The children meekly did as instructed and followed behind Dr. Griffin as she led them to the Lower School dormitory. The rooms had been set up for two occupants during the school year, but with no other students there, each child had their own. The rooms were clustered around a small lounge space that had a window with a view of the campus grounds. Jayla’s room had its own bathroom; there were two at the end of the hall for the others to share. Dr. Griffin indicated she would be staying at the other end of the hall, as much removed from the children as she could be while still providing a nominal adult presence.

Their luggage had found its way to their assigned rooms. Dana wasn’t sure whether to be happy or sad that she didn’t have a roommate. It was always hard to get to sleep in a strange new place, and it would have been nice, in some ways, to share a room with Noah.

“The buildings on campus are security-controlled by a biometric access system, so you will need to register your biodata so you can access your assigned areas.” Dr. Griffin produced a tablet screen and did something on it while having each child touch the handle of his or her door. After that, the lock would only open for the room’s registered occupant and, presumably, a supervising adult. "There is a nine p.m. curfew, after which everyone shall be in their own rooms. And there shall be no noise after curfew!" She disappeared into her room and shut the door.

Dana left her door open while she unpacked. Noah popped his head in, held out the sample case with the newly caught glass lizard in it, and said, "Want to come back to the lab with me? I'm going to set something up to monitor ML4 right now before it disappears on me."

"ML4?"

"Yeah. You said you had captured one of these lizards before and it disappeared. Then we caught one on your deck. Then we got one at the daycare. That makes this the fourth lizard we've tried to hold in captivity. Mystery Lizard Four."

"Got it. So do you want to help me set up a camera?"

"Sure. I think we have time to do that and be back in our rooms before nine. Dr. Griffin sounded serious about that curfew time."

They hurried to the lab, and got busy settling up a small terrarium and a webcam, which they hooked up to one of the computers. ML4 was now under constant video surveillance. If it pulled a disappearing act like its friends had, they would be able to see exactly what happened.

Dr. Griffin was knocking on each child's door for bed check as Dana and Noah got back to the dorms. She nodded as they scooted passed her, saying, "Just in time."

Noah went into his room. As Dana was walking towards hers, the door at the end of the hall opened and a tall, heavyset, very very white man appeared. "Ah, there you are, Gretta," he said. "What fun, to sleep in the dormitory!"

Dana looked from the albino man to the albino woman and said, "Is that your brother?"

The man put out a blue-veined hand and said, "Hello there, I am Mr. White."

Dana, at a loss for words, slowly raised her own hand to take the man's, but his hand was slapped away by Dr. Griffin before it reached Dana.

"No, he is not," Dr. Griffin said. "Not my brother, and not Mr. White." The very white not-Mr.-White was tee-hee-heeing at his joke, his somewhat portly belly shaking underneath his white nightshirt, looking like Santa Claus might if he got locked out of his house at the North Pole and froze to death. "This is my *husband*, Professor Hanz Hetzler. He is also a teacher here at the Cavor School."

"Oh," said Dana, "I thought...." She bit her lip, not wanting to bring up the two professors' matching extreme pigmentation deficiencies. "You don't have the same last name?"

Professor Hetzler stopped chortling and said, "Oh my dear, naming norms are quite arbitrary and particularly subject to the whims of changing trends — for example, before 1900, alliterative[47] names were not common practice, but then suddenly we started seeing them all at once in many places so much so that by 1930 it was a very common custom, and if we combine that with the cultural changes in marriage and divorce rates and their corresponding societal changes in acceptable patriarchal name conventions, we find that many women prefer not to change their name, as they are attached to the sonority[48] of it and their identity, particularly in a professional setting. Gretta and I are not members of a culture in which the women take the man's last name when they marry, and sociological research on personal name use and function has been remarkably sparse, given that —"

"Darling, that is enough," said Dr. Griffin, giving her husband a surprisingly soft smile that could almost be construed as loving. "Now is not the time for a lecture on social naming customs."

"What culture are you from?" Dana asked. Then, before she could stop herself, she asked her real question. "Is everyone so pale where you're from?"

---

[47] *Alliteration* means "several words have the same first sounds," like "bouncing baby boy" or "Dana Dash."

[48] Related to the word *sonar*, which is a device used to detect things by the sound they make. *Sonority* means you like the sound of something, like the sound of your own name.

"Oh," laughed Professor Hetzler, "We were not born this — "

His wife clapped her hands, effectively cutting him off, and announced, "It is now nine p.m. To your rooms!"

That night, Dana drifted off to sleep wondering how exactly it was that someone could become an albino later in life.

CHAPTER 10 — LIZARD LOUNGE

Days passed while Dana and Noah fell into a regular schedule of working on their respective parts of the biosphere project. The aluminum bars arrived, as did the bio-engineering stuff Noah needed. Dana spent time cutting and welding the framework of the Beetle, while he set up his tanks and began building a better breed of bacteria. Life became almost mundane with their ongoing routine. Dr. Cavor stopped by occasionally to check on their progress, but not every day. Days would pass without an adult in sight.

Even their meals seemed to appear out of thin air, with plates of food set out at the appropriate meal times. Dana had no idea if it was done by campus staff or by Dr. Griffin. If it had been magically prepared by elves, Dana wouldn't have been completely surprised. It was always a strange assortment of dishes, with no two plates containing exactly the same meal or beverage, and it didn't seem to correlate to the standard breakfast-lunch-dinner fare. Soup, mashed potatoes, and soda appeared at breakfast along with the pancakes; cereal and orange juice appeared at dinner along with macaroni and cheese. A pile of snacks was always in the center of the table: apples and nuts and candy bars and once, a pack of antacids. A bottle of children's gummy vitamins appeared once a day.

All the children continued to have their meals together at the round table, and gave up trying to make sense of the food system, instead trading and sharing amongst themselves according to their moods and tastes. This was made awkward by the fact that Dana and Lemma pointedly refused to talk to each other at all. Dana reduced this friction somewhat by almost never eating lunch in the cafeteria — she would put two assortments of food on one tray and carry them back down to Jayla's lab. This way, Jayla wouldn't have to motor back up the hill more than once a day.

Dana enjoyed the break from thinking about the biosphere project while she and Jayla geeked out over ham radio stuff. Dana was helping Jayla build a two-meter wavelength[49] antenna in an attempt to establish the Earth-Moon-Earth connection with Jayla's friend across the Atlantic. The idea that they might actually soon be talking to London by bouncing radio signals off the Moon seemed totally wild. Dana was really looking forward to the first test of the antenna, and she breezed through the ham radio licensing exam to get her own callsign. Jayla even loaned Dana one of her fanciest handsets, which they used to chat back and forth across the school campus.

Through all of this, the little glass lizard named ML4 stubbornly refused to disappear in a puff of smoke. It was almost as if it knew that it was on camera and didn't want to reveal its secrets. Noah tried to feed it various things, but it wouldn't eat, only occasionally drink water, and it had become quite lethargic.

Dana didn't want to see the lizard die in captivity and finally decided to talk to Noah about it.

"Maybe we should let the poor little critter go," Dana said. "Watching it die doesn't seem like fun."

"Maybe not," said Noah. "But it's important to finish the experiment. Maybe even catch another one and replicate our results. It's always a good idea to see multiple results before we draw any conclusions."

"I don't like it, if the only conclusion is that we killed it."

"Research animals often have to be killed. That's a part of science."

"That's just sad. And it's not the kind of science I want to do. How about we turn letting it go into an even better experiment? Jayla loaned me some really tiny radio transmitters to play with. What if we could attach a small transmitter to ML4 and set it to ping periodically? We could set up a few listening stations and coordinate the data into a tracking map. Find out where it goes."

---

[49] *Wavelength* means "the distance the signal travels in one cycle of it vibrating back and forth." Two meters is about the same as a really tall person.

Noah's interest was piqued. "That does sound interesting."

"Ok. It's a deal."

Dana assembled the tiny transmitter pack with a little harness made from wire and a zip tie. She took the lizard to the door of their office and set it down.

"Wait!" said Noah. "What about the tracker?"

"I know the frequency. We'll stop in at Jayla's. She has a direction finder. I'm sure we can find it again."

"Can we get that first and make sure it works?"

"Oops. Too late. There it goes!"

The lizard's lethargy had been a ruse. Seeing freedom, it sprang back to life. And it certainly looked healthy as it scurried quickly under some nearby bushes, not slowed at all by the tiny transmitter backpack.

"All right," said Noah. "Let's go find Jayla and see if we can borrow her direction finder."

They walked down the sidewalk in front of the to the door of Lab One. They knocked and Jayla called out, "Come in!"

"Hey Jayla," said Dana as she entered. "Can we use that radio source direction finder you built, so we can track a mysterious lizard that has one of your tiny RF transmitters attached to it?"

"If I had a nickel for every time someone asked me that," said Jayla, "I'd have five cents now." She laughed." My grandma always says that to me. Sure, I'll get it."

Jayla produced a box with four small antennae at the corners and a ring of lights on the top. She flipped a power switch and turned a dial and the box started to beep at regular intervals. One of the lights in the ring pulsed with each beep. "Well," Jayla said, "The signal is strong. It can't be very far away."

"We just let it go," said Dana. "It was moving fast, but it can't have gotten far."

Jayla followed the signal away from the lab door, down the length of the office, towards the bathroom, then stopped. "Definitely not very far," she said. "I'm getting a lot of change of direction angle, just from moving this little Roanoke Doppler[50] a short distance." She pointed to the wall of her lab. "Your lizard may be on the other side of this wall."

"That's Lab Zero," said Noah, "Professor Moreau's lab!"

"His is the last lab in this building," said Dana. "Maybe the lizard is on the other side of his lab, outside the building."

"Maybe," said Jayla, "but I don't think so. The other point of the triangle, based on directional readings, isn't very far through that wall. Somewhere towards the back of the building."

"Easy enough to check," said Dana. "Here, let me have the direction finder." She went back outside holding the beeping box. She walked down to the end of the long building, past the doors to Lab Zero, and stood at the corner to get a bearing on the signal. The next time the box beeped, the light on the LED ring pointed right at the corner of the building — without a doubt, inside Lab Zero.

Tall bushes mostly obscured the outside wall of the lab from view, and it could only be seen standing exactly at the corner like Dana was standing. As Dana looked at the wall, wondering how the mystery lizard got on the other side of it, she noticed the wall looked... strange. Wiggly, like walls certainly shouldn't be. It looked like it was moving! Looking more closely, she realized the wall wasn't moving, but was in fact crawling with mystery lizards, which had taken on the same whiteness as the building's exterior.

Noah had followed her outside and caught up to her. He followed the direction of her gaze. "Ohhhh," he gasped. "Jackpot! Look at these beauties!"

The little lizards were ebbing and flowing, like a pale scaly tide, through a high vent in the wall — crawling with sticky feet and leaping to and from the hedge. Dana shuddered involuntarily.

---

[50] *Roanoke Doppler* uses differences in signal across several closely spaced antennas to find direction.

"The direction finder certainly indicates that ML4 is in that lab," said Dana. "And when we let it go, it went directly to that vent."

"Maybe it's some sort of colony!" said Noah. "Look at all the different shapes! I need to go back to our lab for some specimen boxes! This is so exciting!"

"Forget catching them," said Dana. "What I want to know is where they're going."

She marched back to Jayla's lab. Noah followed, almost giddy with excitement.

"What's the plan?" he asked, hopping up and down as he walked. Dana wasn't sure she had ever seen him so animated before.

"Jayla and Professor Moreau's labs are a mirror image of ours and Dr. Cavor's. If we have a hidden hallway to Dr. Cavor's lab, I'm certain Jayla has one to Professor Moreau's lab too."

"Ooooh right right right," Noah said, still skipping along beside Dana's determined stride.

They returned to Jayla's office and confirmed her suspicions about where the lizard had gone. Dana hesitated for just a moment, wondering if she should tell Jayla about the secret passage to Dr. Cavor's lab, and decided that if she could trust anybody at this school besides Noah, it was Jayla. Jayla was as excited about the secret passages as Noah was about the lizards, and they all started pulling books off the bookshelf in the corner. Sure enough, there was another hidden door. Noah tried the handle. With a click, it opened to his touch. "It's unlocked, just like the one we found in our lab," he said.

"Are we going in?" asked Jayla.

"Oh, I think we have to," said Noah. "These glass lizards are so weird, I don't think I could stop myself."

Dana laughed and said, "Well, since Noah is usually *my* voice of reason and restraint, it looks like we're going to do some trespassing. Coming, Jayla?"

"Why not? I don't get many opportunities to do this kind of thing. Will you be my cellmate if we go to prison?"

"I call top bunk!" said Dana. She was worried for a moment that this might be taken as a comment on Jayla's limited physical abilities but Jayla laughed and agreed that would probably be the best arrangement.

Noah was already halfway down the passage. Dana and Jayla moved quickly to catch up. He reached the door at the other end and put his ear to up to it, listening. He waited for several seconds like normal-patient-Noah would, but unable to stop himself any longer, he opened the door a tiny crack and peered through. He gasped softly.

"What do you see?" whispered Dana.

"What's that smell?" hissed Jayla, wrinkling her nose.

In response, Noah opened the door wider. Dana pressed her head next to his and peeked through the gap, and Jayla ducked her head to look out under his arm.

Professor Moreau's lab was filled almost to the ceiling with aquariums and cages, organized in neat, symmetrical rows. Many of the animals were familiar: mice, rats, guinea pigs, rabbits, monkeys. But some were animals that Dana had certainly never seen before — that she bet *no one* outside of this lab had ever seen before. Lizards with colorful feathers, lemurs with batlike wings, rabbits with large visible canine fangs. Dana made a strangled noise in her throat, and it wasn't because it smelled like a zoo.

There didn't appear to be any human animals. Noah opened the door all the way, and the three children entered the lab.

The animals reacted to the intrusion like a biological security alarm, with a chorus of screeches, hisses, hoots, and howls. The young intruders froze and listened for any human response. When none came, and the animals quieted down, the trio proceeded cautiously deeper into the lab.

Noah stopped to look at every new oddity, but Dana was mission-driven. "Let's find ML4. We can look at all of the rest of this after that if we have time. Jayla, what does your box say?"

Jayla said, "The angles on the direction finder would put the radio source closer to my lab. So the transmitter is in the back of Professor Moreau's office."

"Ok," said Dana, "But the vent those lizards were coming in would be somewhere over there." She pointed to the north wall.

No overhead lights were on, but the illumination coming from some of the animal enclosures was enough for Dana to see a white plastic tube running along the ceiling. The placement was exactly right to connect to the outside vent, and it was large enough for the small lizards to move through it like a highway. She followed the tube across the room, where it disappeared into the wall above Professor Moreau's office door.

The three moved quietly through the darkened lab. The blinds on the office windows were drawn, so Noah again put his ear to the door to listen for signs of occupancy, though it seemed unlikely that someone would have been inside and not heard the din when they entered a moment before. Noah slowly opened the door. Unlike the lab, the light was on. The walls were lined with tables, mirrored against each wall, with neatly stacked piles of papers and a very logically arranged assortment of office supplies. The bookcase on the far wall, Dana noticed, had the books organized in a chromatic symmetry by the colors of their spines, so that the entire wall of books resembled a color wheel.

Above one table was a large bulletin board with a tidy arrangement of items pinned to it. In the middle of this was a map of Long Island[51] and surrounding areas, including New York City to the west and Westchester County to the north, where there was a large yellow pin marking the location of Dash-on-Hudson. But Dana's attention was drawn to the other side.

A small blob of land on the very easterly edge of the map, labeled Plum Island, was encircled in red marker. A red string began there, and went from pin to pin — first to Montauk,[52] then down the length of Long Island, dotted with pins along the way. Each pin in this line had another

---

[51] Part of New York State, the west end of which has the New York City boroughs of Brooklyn and Queens.

[52] The far eastern point of Long Island. It has a very nice and useful lighthouse, the oldest in New York.

red string linking it to a newspaper article, with titles like *Mysterious Burglary Suspect Still at Large* and *String of Strange Sightings Continues.*

The Montauk pin was tethered to an article entitled *More Montauk Monster!* with a close-up photograph of a bizarre-looking carcass washed up on a beach. It had a dog-like body, but without fur, and in place of paws it had long claws that looked like human hands. What remained of its jaw had razor-sharp bottom teeth but what looked like a bird's beak on the top. Written in the margins of the picture was a single question: "Romulus or Remus, halfway through metamorphosis?"

Noah leaned forward and read aloud in a low voice. "The creature that washed up on the beach in Montauk has sparked much speculation and controversy, with some suggesting it was a shell-less sea turtle, a dog or other canid, a sheep, or a rodent of unusual size. Others believe it was an elaborate fake. A few have spun theories about genetic experiments from the nearby Plum Island gone awry. Moses Moreau, head of research at the Department of Homeland Security's Plum Island Animal Disease Center,[53] denies that genetic experiments on animals have ever taken place at his facility."

Dana said, "Those creatures in the other room make me think Professor Moreau might have been fibbing *just* a little. That photo certainly looks like something that could have escaped from that lab."

Jayla still had the RF detector on her lap. "The signal is coming from in there!" she pointed to the bathroom.

Dana and Noah looked at the bathroom door. Noah hesitated. Dana said, "Go on, this is your lizard quest."

"All right," said Noah. "Let's see what you impossible little squamata[54] are up to..."

Noah opened the bathroom door.

---

[53] This is a real place that really denies conducting experiments of this nature and any connection to the *Montauk Monster*, which, well, go ahead and look it up. You know you want to.

[54] Animalia-Chordata-Reptilia-**Squamata** is the scientific classification of the order of lizards within the animal kingdom (and sounds like a good rope skipping rhyme).

The bathroom had a sink and toilet crammed into one corner, so it was still useful as a bathroom. But its bathroom-ness ended there, and the rest of the space was something else entirely. Against the far wall, on a burgundy velvet chaise, a man was lying on his back.

The children froze — not only because they didn't want to be caught trespassing, but also because the man's appearance was alarming. It wasn't his body, which was dressed in a white lab coat over red slacks and a red shirt. It wasn't his feet, which were bare and a little gnarly, but still looked like normal feet. It wasn't even threadbare pink-flowered slippers resting on the floor next to the chaise; they were almost comforting. It was the man's face, which was completely covered with a gas mask.

The gas mask was the kind Dana had seen in the black-and-white sci-fi movies her dad liked: old-fashioned, with round eye pieces made o f dark glass, and a flexible hose snaking from the mouthpiece. A hose for an apparatus such as this would normally run from the mask facepiece into a filter canister, but this hose was much longer. The children, almost in unison, traced the trailing arc of tubing upwards, where it connected to the funnel-shaped bottom of large, only semi-transparent aquarium tank suspended from the ceiling. Inside the tank was a swarming, writhing, churning mass of lizard shapes. Some were coming and going through the white tube lizard highway that ran from the vent outside the building on the far wall of the lab. But there was another lizard road: into and out of the gas mask mouth via the hose.

The three children stood there motionless, staring, not quite believing what they were seeing. And as they watched, they could see the man's exposed throat move as his chest rose and fell with deep, slow breaths. Wriggling. His throat was squirming. He was breathing the lizards in and out. Or maybe he was ingesting and regurgitating them. Dana felt her stomach heave, and she might have thrown up except that it wasn't yet lunchtime and her stomach was empty.

"Nope! Nope! Nope!" Jayla yelled as she threw her wheelchair in reverse so fast that she almost left skid marks on the office floor.

From the notebook of Dana Dash

CHAPTER 11 — STILL LIFE

The three children got out of Professor Moreau's lab office as fast as they could. Jayla zipped swiftly back the way they had come. Dana and Noah were close behind, tripping over each other to get out the office door. Dana threw a glance behind her as she exited the office and saw the man was starting to sit up slowly and stiffly, like Frankenstein's monster, probably disturbed by the commotion of the three children beating a hasty and undignified retreat. Then she was around the corner and out of sight. Jayla was far in front. She was halfway down the connecting passage before Dana and Noah reached it. Her wheelchair must have some sort of overdrive modifications — Dana had never seen one move like that.

Noah was last through the passage door on Professor Moreau's side and closed it tight behind him, and when they were all through the door to Jayla's lab, he slammed that door too. Dana helped him push the shelving back into place. They caught up with Jayla in her office.

"What the holy heck was that?!?" Jayla practically yelled.

"Oh, like we've ever seen something like that before?!" Dana half-yelled back at her.

"Calm down both of you," yelled Noah, and then in a yell-whisper, "Did he see us?"

"I don't know," said Dana, also lowering her voice. "Maybe. I don't think so. But he definitely started to wake up when Jayla freaked out."

"I think that warranted a freakout," said Noah. "I've read about some cutting-edge genetic engineering, and maybe some of the stuff we saw in the lab is possible, but the lizards and the mask... I don't even know what to think about that."

"What do we do now?" asked Jayla. "He definitely knows someone was there."

"Let's get out of here," said Dana decisively. "Jayla, you head up the hill for lunch. Tell anyone you see that I couldn't hang out with you today. Noah and I will go back to the lab and at least pretend to be working. Leave your door unlocked, so if he figures out someone came through your lab, you can plausibly deny it was you."

"Better than any idea I've got," said Jayla. "But let's get together tonight and talk about this. This whole school is so weird."

"Go now," said Noah. "Talk later. He could come in here any minute."

Jayla and Dana both nodded and the three children got moving.

Dana and Noah arrived back in their lab minutes later.

"Ok, um, look busy..." Noah said.

Dana stood there, unsure of what to do with her hands. She locked eyes with Noah and slowly raised her arms out in front of her and waved them around mechanically, pretending to touch things. "Working working working beep blerp blorp," Dana said in a robot voice.

"Be serious for a minute. We need to look busy." But he was smiling now.

"Well, the next thing I was going to do was bug Dr. Cavor again about those windows," Dana said. "Oh! If I'm over at her lab talking to her, I can't have been in Professor Moreau's lab!"

Noah nodded. "Good thinking."

Dana hurried to the lab next door and knocked on the door. No answer. She tried the doorknob; it clicked, then turned. Startled, she let go of the knob.

The door swung slowly, silently open.

"Dr. Cavor?" Dana called out. "Are you there?" Silence answered her. Dana took a step inside and tried again. "Dr. Cavor? I'm here for the lab window plastic we talked about?"

More silence. As her eyes adjusted to the darkness, Dana looked around. The walls were plastered in worn yellowed papers, smeared with inky fingerprint smudges. Amongst the papers were a few large, hand-drawn maps, with smatterings of faded, delicate handwriting. She traced her hand along the wall to a section with diagrams of machines like nothing she'd ever seen before. Dr. Cavor's organizational methodology wasn't obvious to Dana. Maps were intermixed with the diagrams and diagrams were shuffled in with the maps.

Dana supposed that Dr. Cavor could've been simply careless while tacking up her papers, but somehow Dana didn't think so. She was beginning to suspect everything Catherine Cavor did was deliberate, which was something they had in common.

"Dr. Cavor?" Dana called out again as she walked into the lab proper. It was filled from the floor almost to the ceiling with tall shelves, packed with tools and equipment. The shelves didn't merely line the walls; they were stationed like sentries throughout the space, blocking any direct sightlines, and sectioning the room off like a maze. Dana entered, and the shelf-maze opened into a clearing with a lab table, stacked a meter high with odd objects and books and papers and who knew what else. Dana was afraid to touch anything for fear a pile might tip over. She would never be able to put it back exactly the way it had been.

Dana moved deeper into Dr. Cavor's shelf-walled lab-maze, turning right, then left, then right again, until she reached another clearing. Instead of a lab table, there was a large glass case on delicate filigree brass legs. It was a human-sized case that looked very much like a transparent coffin — not just because of the size and shape, but because inside was a human-sized form, covered with a sheet. *A mannequin? A* — Dana gulped involuntarily — *a body?* Whatever it was, it was certainly a human female shape. Dana thought it seemed more like a sleeping princess fairytale than something out of a horror movie. She raised her hand to touch the glass, but a feeling of intense cold stopped her fingertips before they reached the smooth surface. It was a cold colder than anything Dana had ever felt, but localized directly next to the coffin-bed-incubator-Sleeping-Princess-case. For air that frigid, Dana would have expected the glass to be covered in frost, but there was none. She glanced down at her feet and saw a perimeter of frost on the floor all around the coffin-case-thing. It was so cold that moisture was actually condensing out of the air before even touching the glass. Was she looking at a frozen dead body? Or maybe a cryonically frozen body, still alive? She thought about her father's medical bracelet. Could the Cavor School and Dr. Cavor be practicing cryonics?

On either side of the case were matching metal frames, the sort of stand that one might find in a toy shop for displaying large dolls. One of these frames was empty — but one was not. It was occupied by a very life-like robot android, also in the form of an average-sized adult human woman, although average adult human women have faces, and this android had a jumble of wires and circuits nestled within a large hole in its head. Flesh-colored panels on its abdomen and limbs were ajar, and inside were webs of flashing lights and machinery. Both the android, and the empty android stand, had cables running between them and the fairytale-robot-coffin.

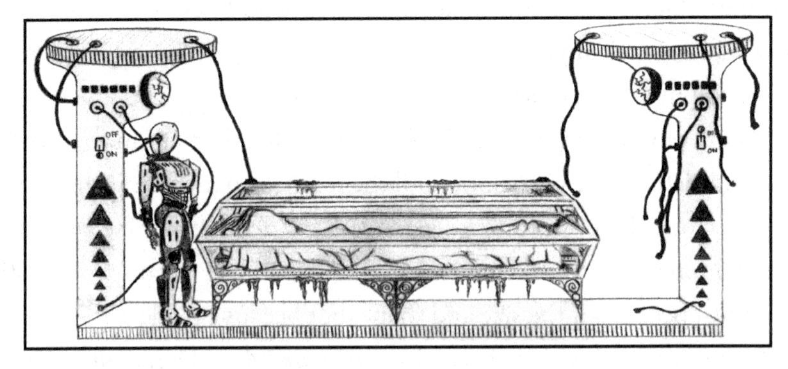

From the notebook of Dana Dash

Dana stood frozen in front of the whole scene, fascinated and dumbstruck and really at a loss for what to do or think. Could something be metal and plastic, yet still be alive? How long she would have stood there she would never know, because out of the corner of her eye she thought she saw a flicker of light. And when she turned to look, the whole bizarre robot-doll-freezer-display-case setup was all but forgotten. There, in another open space along the maze path, Dana saw an old-fashioned writing desk. And on this desk lay an old-fashioned black umbrella.

Dana walked gingerly around Dr. Cavor's mannequin-android-cadaver setup and made her way towards the desk. It was the same black umbrella from Dr. Cavor's office at the manor house, with the same strange blue and orange swirled candy stone that was both sparkling and lifeless, set in an ebony double helix handle. Dana barely registered that behind the desk, against a wall, were the large pieces of plexiglass she had come to find. She only had eyes for the umbrella.

Dana started to reach out for the stone but then hesitated. What exactly *had* happened the last time? She remembered darkness and garbage — a dark figure and a horde of shadows, the sense of being trapped in someone else's body — but the details were gone, like a dream she couldn't recall. What she did remember was the craving to touch it again. She wanted to repeat the experiment, because the ability to replicate results is the foundation of good science.

Trembling a little, half with excitement and half with fear, Dana sat down on a wooden chair in front of the desk. She took a deep breath, and another, and then a third, willing herself to be calm and focused. As she exhaled the third breath, she reached out her right hand. She touched the stone gingerly with her fingertip and...

Absolutely nothing happened. Or she *thought* nothing happened. She was still sitting on a chair, in front of the same desk. But when she looked at her right hand, it was resting on the desk. The umbrella was gone. The contents of the towering maze-shelves surrounding the desk had changed: where before there had been antiques and odd machinery, the shelves were now crammed with books.

The desk in front of her also had a set of leather-bound volumes all along the back — when she sat down, there had been a row of decorative paperweights where the books now sat. They had no titles, but instead were ordered by golden numbers on the spines. Thinking of the dodecahedron she was assembling next door, she grabbed the book labeled with the number twelve. She rifled randomly to a spot in the middle and began to read the handwritten page.

> There is evidence that the Xenos have been with us for a very long time. They have scientific knowledge verging on the magical, far beyond those of humans, and could likely rule or destroy us at their whim. Despite this, they seem to prefer being hidden, appearing only as rumors and legends, fables of faerie folk and fiends, demons and demigods.
>
> Perhaps, they are not hiding from us but from each other. There is evidence of at least two factions of Xenos that have, at various times, been at war with each other. Moses believes that the creature who seems drawn to the Moonstones is a relic from one of these wars — a constructed weapon of some sort. He has further theorized that it is trying to reconstruct a body around itself, which explains why it manifests as a collection of different objects. We have agreed it is best to keep these stones out of its hands, and are continuing to study the stones we have gathered for clues to their origins and to the secrets of the Xeno technology behind their creation.
>
> Each of these stone is an irregular shape, with no sharp edges. Some seem to fit like puzzle pieces. The stone in my umbrella, which I have been able to use for memory storage, among other things, was originally two stones that, when pressed together, joined seamlessly before my eyes, their colors melding together as if they had never been apart.

"Hey Dana? Did you find the plastic?" Noah called as he came in behind her.

Startled, Dana closed the journal with a snap and quickly re-shelved it in its correct place.

"Yeah. It's over there by the wall," she said. "I was just..."

She glanced in Noah's direction, then turned back to see that the books were gone and the row of paperweights was back. Her right hand rested on the desk, with her outstretched finger hovering above the umbrella handle, no longer touching the stone. Her hand was where it had been the whole time. She had never actually picked up and read a book.

"You were just snooping!" Noah laughed. "Totally understandable. Look at all this cool stuff!" Noah walked further into the shelf-wall-lab-maze. "Whoa! Like, what is this thing here?"

"Wait, come back for a minute," Dana said. "I want you to test something."

Noah poked his head around the corner of a shelf-wall. "Sure. What is it?"

"It's Dr. Cavor's umbrella. The one from her office. I want you to touch it."

"Oh, right. You said it did something funny to you. What does it feel like?"

"I don't want to contaminate the results by telling you anything else. Try it and tell me what happens."

"Ok." Noah walked over to the umbrella and touched it.

"No, I mean touch the handle."

Noah grasped the smooth double helix-shaped wood, and shrugged his shoulders. Dana sighed.

"The *stone*. Touch the stone in the handle."

"Oh." He let go of the handle, and, stretching out a single finger towards the blue-orange swirl, touched it lightly. Dana studied his face closely, watching for any change in his expression. He looked back at her intently.

"I don't feel anything. How long does it take? Should I press harder?"

Dana shook her head, and Noah released his touch.

"Nothing happened?" she asked. "You didn't see anything?"

"Nothing but you watching me. But come look at this thing over here!" He walked back around the bookshelf.

Dana sat in the chair for a moment, chewing her lip. She could conclusively say now that *something* had happened to her both times, even if she didn't know *what* had happened. The journals said something about memory storage and the stone. Why would she be able to see these visions, but not Noah?

"Dana!" Noah called out from the other side of a bookshelf. Still thinking, Dana stood up and followed Noah to see what he had found and was so excited about. He had lifted up the corner of a dusty tarp covering a table. The source of his excitement was the source of a faint blue glow and a barely audible hum, and it temporarily shelved the mystery of the umbrella stone, replaced with the mystery of this strange contraption. It was smallish, squareish, made of black and shiny metal, about the size of a breadbox. A cylinder ran lengthwise through the center, carrying a small conveyor belt through it from one side of the machine to the other. Dana impulsively flipped a switch, and with a low whir, the belt slowly started to move. It now appeared not to be one continuous belt that entered one side of the cylinder and came out the other but in fact two belts, running in different directions, like two moving walkways pushed together so that people walking on them would run into each other in the middle.

"Um, maybe we shouldn't be randomly pushing buttons until we know what this does," said Noah.

"It wasn't random; it was clearly a switch to start the belts moving," said Dana. "The main controls are here on the front."

"Um," said Noah, "Then maybe we shouldn't be messing with those until we know what this thing does."

"Obviously you're meant to put something on these belts and have it feed to the center. But where does it... hey, see those paperweights on the desk over there?" Dana pointed to where the umbrella lay. "Hand me one."

"Um," said Noah, "maybe we shouldn't be sticking random things into there." But he retrieved a glass paperweight with a red rose encased inside, and put it into Dana's outstretched hand.

Dana put the glass rose on one of the belts, and they watched it slowly chug along into the interior of the strange machine. As it disappeared from view, Dana wondered how a glassblower was able to trap a flower inside molten glass without burning it.

From the notebook of Dana Dash

"Oh!" they both said as the paperweight emerged on the other side. It seemed unchanged, but when it exited the machine, it was riding, not on top of the belt where they had placed it, but on the bottom.

"So it *is* just one belt," said Dana, "It must twist somehow in the middle."

"Is the belt sticky or something?" asked Noah. "How is the paperweight staying on the bottom like that? It seemed pretty heavy."

Dana watched the glass-enclosed flower, affixed to the bottom of the belt, whir along and start up around the curve. She expected it would continue, stuck to the belt, and head back through the machine the way it had come. But as it slipped around the edge of the curve from the bottom side of the belt to the top, it wobbled, and Dana expected it to fall. And it did.

But it fell up!

The pretty paperweight plummeted to the ceiling of the lab. Dana's head snapped back, eyes tracking the rapidly rising red rose until it smashed into the ceiling, where it shattered into a gajillion sparkling small pieces of glass. The explosion of shards bloomed outward and only temporarily downward from the impact, before curving back to the ceiling, some of it skittering quite a long way across the underside of the roof.

Dana and Noah both instinctively ducked and shied away from the anticipated shower of glass. When it didn't come, they both slowly looked up at the ceiling and stared, fish-faced, at the distinctive shatter pattern of dropped glass, which one must occasionally clean from a floor, but which one almost never, ever, needs to clean off a ceiling.

"Hello?" An unfamiliar male voice called out. "Catherine? The door was open..." The two children froze as footsteps moved through the office towards the door to the laboratory shelf maze.

Then the sound of Dr. Cavor's voice carried across the lab, "Moses! How did you get in here?"

Moses — the voice belonged to Professor Moreau!

"Catherine, there you are. I was just looking for you. The door was open."

"Was it really... It is certainly unlike me to leave it unlocked."

"Not unlocked — wide open. I assumed you were inside." Professor Moreau sounded defensive.

Noah was waving wildly to Dana, trying to get her attention. She was looking, past the glass coffin thing, at the door to the office. Dr. Cavor would be coming through that door any moment, to check for other uninvited guests. Dana looked at Noah, who mouthed the words, "*SECRET PASSAGE*," and pointed to the back corner of the lab, behind the mysterious gravity-reversing machine. Sure enough, there was a door, like the one in their lab, only this one wasn't hidden.

Dana vigorously nodded and was about to head towards it but, on impulse, she flipped the switch to turn off the conveyor belt and picked up the machine. It seemed odd that something which had moments before shattered her understanding of the governing laws of the physical world was so small and light that she could tuck it under her arm as easily as a box of new shoes. With her free hand she hastily rearranged the tarp so it was balled up under itself and, at least from the front, maintained a passing resemblance of still containing its recently-former occupant. Then she made a beeline towards the passage door.

The door opened as easily to Dana's touch as the one on their side had done. She waved Noah past her and shut the door, quietly as possible, just as she heard Dr. Cavor enter the lab area.

"Wait a moment," Dr. Cavor said.

Dana's heart raced. Had she been seen?

Dr. Cavor continued. "I need to examine my lab, to make certain no one else wandered in and that nothing has been disturbed."

"That you should," said Professor Moreau. "I had some uninvited guests in my lab earlier. I did not see them, but I heard them." Dana thought he certainly sounded normal for someone she now assumed was made entirely of lizards. Which, she supposed, was not worse than snakes and snails and puppydog tails.

Dana reached out in the darkness and found Noah's arm. She squeezed and then pushed gently and they sneaked off down the passage, as quickly but quietly as two students stealing a magical mystery machine from their teacher could possibly be expected to sneak. They opened the far door, pushed the bookshelf forward, scrambled out, and closed the door behind them. As they stood there looking at each other, wondering what to do next, Dana had an inspiration.

She set the machine out of view under a lab table and ran quickly out of their lab, outside, and over to the open exterior door to Dr. Cavor's office in Lab Five. Dana took a deep breath and walked in, saying, "Hey Dr. Cavor, I need..." and stopped short at the sight of a white-haired gentleman in a white lab coat over a red shirt and pants. She knew this must be the man who had been wearing the gas mask, but his face looked familiar, even though she hadn't seen him in either his lab or Dr. Cavor's before this moment. *Probably saw his picture on the staff directory wall at the manor house,* she told herself. *Or at the welcome ceremony.* "Oh," she said, "I'm looking for Dr. Cavor. Is she here?"

"Yes. She is in her lab. Catherine, you have another visitor!" he called out to her.

"My private work areas have become adjuncts of Grand Central Terminal," said an obviously annoyed Dr. Cavor, returning from the lab area. "Moses, this is Dana Dash, one of our STEM contest finalists. Delta

Discipline. Dana, this is Professor Moses Moreau. He is the Master of Science, and the professor who teaches our advanced bio-sciences classes."

"It is a pleasure to meet you, Dana." He smiled in a way that Dana could only describe as "boyish" as he extended a hand.

"Yes, likewise, Professor," said Dana, shaking his hand. It felt like a perfectly normal hand. Not at all like a bunch of tiny lizards wearing a man suit. *He's a normal everyday guy with a moderate lizard-breathing habit. No big deal, right? Right.*

Dana let go of the non-lizard-filled hand and resisted the urge to wipe it on her jeans. She turned to Dr. Cavor. "I'm sorry to bother you, but I saw the door was open, and I've been trying to contact you about that polycarbonate sheeting you told me about."

"Ah, yes," said Dr. Cavor. "I will have them in your lab tomorrow morning. Now, unless there is anything urgent, Professor Moreau and I need to discuss some matters."

"No, thank you very much. Um, bye..."

"Good day," said Dr. Cavor and Professor Moreau in unison, and Dana walked out of Dr. Cavor's office and returned to her own lab at what she very much hoped was a normal and nonchalant pace.

"Quick thinking!" whispered Noah, who had been lurking inside their own office door, listening to the conversation. "But do you really think it was a good idea to take her machine?"

"If she notices it's missing or looks up and sees the broken paperweight, she will probably assume someone came through the front door she left open. Anyone could have done that. Here, help me push this shelf back where it was, so even if she does think of the other door, she won't have any reason to believe we had anything to do with it."

As they moved the shelf back, Noah said, "You don't feel bad about stealing it?"

"Let's call it 'liberating for educational use by the people,'" Dana said.

"Yeah, that's you, a regular Robin Hood of Cavor Forest."

"I'll put it back," Dana shrugged. "But not before we see that machine work again. We need to repeat the experiment! For science."

Noah nodded. "For science you'll smash another paperweight?"

"Maybe. Why not? But just think — gravity-reversing! I can't wait to see what we gizmos we invent!"

CHAPTER 12 — FATHER OF INVENTION

That evening they had a sit-down in Jayla's dorm room after dinner. Dana and Noah had decided not to tell Jayla about the strange, albeit fascinating, things in Cavor's lab. Jayla was already concerned enough about the amount of strange in Professor Moreau's lab.

"Do you think its even safe for us to stay here?" Jayla asked, tugging on one of her braids.

"I don't think it's really a threat to us," said Dana. "It's still a regular school, and has been for years. Kids are here all year long and don't get turned into lizard-breathing monsters."

"That we know of," mumbled Jayla, still yanking on a braid.

Dana ignored her. "Besides, isn't it exciting? The science they've got here is more advanced than anything I've even *heard* about. I mean, I always thought adults did weird secret stuff behind closed doors, but I had no idea it could be this *amazing*!"

"But the key point is that it is a secret," said Jayla. "If they figure out we know something — I don't want to find out what they would do. Maybe I should tell my grandmother and see what she thinks."

"I can't believe you're being so — "

Noah intervened. "Dana, Jayla has a point. That stuff in Professor Moreau's lab was seriously freaky, and something we're definitely not supposed to have seen. But Jayla, Dana is right too. We want to go to this school *because* of the advanced science. Should we change our minds because the science is *too* advanced? For me — I want to know everything about whatever weird genetic engineering Professor Moreau is doing. I'm just not going to learn that stuff at Hillhurst Middle School. Or probably anywhere else."

"Right!" agreed Dana. "Think about it, Jayla. Dr. Taxidiótis probably has equally advanced engineering to teach you. I thought the crazy biology stuff was creepy and gross, but what if they have crazy physics stuff here too?" Dana wished they hadn't decided to keep the antigravity a secret from Jayla. That was something Dana was pretty sure Jayla would think was as cool as the lizards had been disgusting.

Jayla continued to tug on her braid, but said nothing for several moments. "I guess I see your point. It's still a school. And I want to be here to learn the most advanced engineering they'll teach me. I just hope you're not wrong about it being safe. My grandmother... I guess I want to stay, and I don't want to give her anything to worry about. She's already..."

Jayla stopped, and let go of her braid. "All right. They have their secrets, and we'll have ours. I won't tell my grandma. For now."

"I think that's fair. For now." Dana put out her hand to shake on it. Jayla rolled her eyes and laughed.

The next day, Dana and Noah were back at work on their project, trying to pretend that nothing unusual had happened. "It's going to be a tough week," Dana said from inside the Beetle, "but I think we can manage it."

The polycarbonate windows had arrived early that morning from Dr. Cavor's lab and Dana had started cutting, fitting, and sealing the plastic into the welded Beetle framework. Noah worked with his bacteria tanks while Dana put on the Beetle's windows. A Beetle-heart, minus the legs, was made the same way, and then braced into the center of the larger framework with aluminum struts. The whole structure was mounted on five legs, on a steel base plate. The whole thing stood about three and a half meters[55] high.

There was an airlock chamber at the bottom, with both an outer and inner airlock door. The entrance to the Beetle was through this airlock, with clearance beneath it under a meter. An adult would have to duck and wriggle into the structure, but for children, it wasn't difficult..

"We won't get claustrophobic with all these windows," said Noah,

---

[55] Approximately 11.5 feet.

climbing through the airlock and standing up right above it where there was no upper or lower plywood decking to allow access between decks. He climbed up to join Dana on the small upper deck and added, "We should spend most of our time sitting or lying down anyway, to conserve oxygen."

"Right," said Dana. "I'm going to string up a couple of sleeping hammocks, from these struts in the ceiling of the lower deck."

"Cool," said Noah. "Hammock-ing! It'll be an adventure."

"Yep. I couldn't do much more with that space anyway. The airlock takes up most of the bottom half of the Beetle. That lower decking is hardly big enough to even sit. It's useful for storing long things, though. I stashed some spare aluminum tubes and my telescope underneath." With the telescope, Dana had also stored the strange medieval sextant device that Ms. Astrulabi had given her. She hadn't had much spare time lately, but But in the week trapped in the Beetle, she would have almost nothing else to do but stargazing. She was almost looking forward to it.

"You wanna beta-test the hammocks tonight?" asked Noah. "It's worth trying it out before we lock ourselves in."'"

"Good thinking," agreed Dana. "I bet Dr. Griffin won't even notice."

"Yeah," said Noah, nodding. "She stopped doing bed checks after the first night. I haven't seen her or Professor Hetzler in over a week. Except for the food that appears, they hardly seem concerned about us at all." He paused. "Not sure if I think it's a good thing or not. This school really operates by a different set of rules than other schools."

"Well, I like it this way. Means we can work on our project and not worry about explaining ourselves to adults all the time. Speaking of working, I think you'll like this — the controls for the inner shutters and grow lights are working. The only issue has been that the little motors I'm using only move in one direction. But I solved that by putting weak springs on the shutters so they return to default position when the motors are turned off.

Also, the outside shutters will have photovoltaic[56] material on them to turn each shutter into a solar panel."

"Cool. I'll be glad to have some privacy from people staring at us in this people-aquarium all week. And also be able to close the shades and sleep while the grow lights are on for the cyanobacteria. Or past six a.m. when the sun rises," said Noah. "Those shutter springs sound really efficient.

"Seems doable. Speaking of cyanobacteria, how is the genetic modification coming?"

"Great!" said Noah. "I isolated the gene sequences that would increase oxygen production, then used the CRISPR to splice them into the bacteria's genome. So now I've amplified the genetic modifications."

"You explained how the CRISPR thing works, and I sort of get it," said Dana, "but what do you mean by 'amplify the genetic modifications'?"

"I mean I need make sure most all of the bacteria we use has been modified, even though the changes I make will only reach a small portion of it at first. You know how evolution works, right?" asked Noah.

Dana tried not to be offended by the question. "Of course."

"Right. So in evolution, we talk about natural selection. Things with traits that are better suited to their natural environment will survive. That's being 'naturally selected.' And that's what we're doing, except it's an 'artificial selection' since we're growing this all in tanks and not the ocean. So now that I've changed some of the bacteria's traits, we need to select for only those bacteria to grow."

"So, we need to put it in an environment where only the best bacteria survive? How do we do that?" asked Dana

"Well," said Noah, "I cheated. Not exactly cheated; it's a pretty common technique. I spliced in a couple extra genes, including antibiotic resistance. So when I expose all the bacteria to an antibiotic, the unmodified bacteria dies, and the good stuff survives."

---

[56] *Photo* means "light." *Voltaic* means "producing electricity (volts)." Each shutter is a solar panel.

"Cool!" said Dana, "That's really slick. But how do you know it worked?"

"Oh, that's the best part. The other gene I added was for bioluminescence. So if it glows, we'll know it was altered. Come here and look at this this." Noah climbed out of the Beetle and walked over to turn off the lights.

It took a moment for Dana's eyes to adjust as she climbed out of the Beetle too but, when they did, she could see that Noah's tanks were giving off an eerie, pale blue green light.

"Ohhh that's beautiful," said Dana. "Your magic-algae-oxygen tanks look like actual magic now! I'm awarding you a hundred mad-scientist points for the glowing bacteria tanks"

"Thanks," said Noah. "I think it adds a certain something."

"Totally. I can't wait to see what the contest judges think."

"We'll see," said Noah. "Now all we have to do now is give the cyanobacteria the light and carbon dioxide it needs. I know how much CO_2 I'm giving it, but what I need is a sensor, so we can tell how much oxygen we're getting out."

"Sounds like my kind of project," said Dana. "I can add an oxygen sensor and a display gauge to my Beetle control panel. Oh, and I'll have it sound a warning alarm if oxygen gets too low in our habitat. I'll have time. I'm done cutting and installing the windows. The Beetle frame is done, I just need to seal them with gasket material to make them airtight."

"And pressurized," said Noah. "That has to make it harder."

"Actually, I think the pressure may make it easier to get an airtight seal," said Dana. "The frame will hold the windows in place, then the internal pressure does most of the work of pushing the gaskets into a tight seal. Same way the zipper seals on our EVA suit will work."

"Sounds good enough for our purposes," said Noah, "but I'd be terrified if we were actually going into space and you were telling me that the only thing holding air inside was the air itself trying to escape."

They both had a nice long laugh at the notion of going into space in a glued-together science project contraption.

That night they slept in the Beetle. Dana woke up, a little stiff, to the sound of her phone ringing from a hanging pouch next to the hammock. She checked the caller ID and quickly answered.

"Dad?" she croaked.

"Hey Bug! Did I wake you?"

"No, no, I'm up," she said, stifling a yawn. "Is everything ok?"

"Your mom and I are being spontaneous! We're going on a trip! I remember you saying animals were allowed at the Cavor School. Could Bosco and Chip stay with you and Noah while we're gone?"

"Yeah, animals have to be allowed on campus for scientific experimentation and stuff. Let me see what Noah thinks. Hey Noah. Noah!"

"Hrmph... wazz?" mumbled Noah sleepily from his hammock.

"How would you feel if Bosco and Chip joined us? My parents want to be spontaneous, but somebody needs to look after the critters."

"I'm ok with having the canidae[57] aboard," said Noah, "Chip can stay in the lab, but what about Bosco?"

"Pretty sure nobody will notice. And even if they did, I think they'd pretend they hadn't. Most adults would rather ignore a problem than deal with it. Plus, we can say it's part of our project."

Noah nodded somewhat sleepily at Dana's logic.

"Ok. Come drop them off." Dana said to her father. "Where are you going?"

"Into the wilderness! Camping and maybe some spelunking with Noah's parents," her father said.

"Sounds really fun. See you when you get here."

Bosco and Chip were dropped off later that day, along with enough food for both Chip and Bosco for the week. Her father helped Noah position

[57] *Canidae* is the biological family of animals that includes both dogs and foxes. Bosco is Animalia-Chordata-Mammalia-Carnivora-**Canidae**-Lupus-Familiaris (domesticated dog) while Chip is Animalia-Chordata-Mammalia-Carnivora-**Canidae**-Vulpes-Vulpes (red fox).

Chip's aquarium inside the inner dodecahedron Beetle-heart, among the algae tanks, while Dana proudly showed off the features of the frame and shutters to her mom. Bosco was tied to one of its legs. He was unhappy about this arrangement. Chip registered no complaints, but remained a fluffy, slumbering red curl. Dana's parents wished them luck on their week-long camping adventure and waved goodbye as they drove off.

The Beetle was nearly ready, but the final preparations had to wait for Noah's bacteria to reach full capacity in all of the tanks they would be using. In the meantime, the gravity-reversing machine beckoned irresistibly. The unstated goal was for them to be skimming along the ground on hoverboards, or cruising through the sky on a flying silver surfboard. But the two children approached their initial experimentation carefully and methodically. The first experiment was a simple test of how much "lift" an object had once it was passed through the machine.

"We need some weights," Dana said.

"No problem. Our lab came with some standard equipment, including a bunch of scales and calibration weights." He went to one of the shelves and pulled down a precision scale from the top shelf and an obviously heavy case from the bottom shelf. The case contained cylindrical weights, ranging from tiny ones that would be hard not to lose to great big ones that would feel very hard if dropped on an unprotected toe.

"Great," said Dana. "Now, we need exact duplicates of these. Let me see what I can borrow from the neighbors."

Dana tried Jayla first, but she was using her weights for her cybernetic interface project, calibrating the lift capacity of a robotic arm. There was no way Dana was going to ask a favor from Lemma, or even her partner Prima, so that left the math kids. Toma and Tevi said no problem; they weren't planning to use theirs. Dana brought the weights and scale back to her lab, and they got to work playing with the mysterious magical machine.

They set up a work table. On the top of the table was one scale and set of weights — regular weights, with positive gravity. The negative gravity weights and small scale, having passed through the gravity-reversing machine, were on the underside of the table. As Noah arranged the weights on the bottom to mirror the positions of the weights on the top, he noticed something and called to Dana. “Look at this! Look at the writing on these weights.”

Dana got down on the floor to look at the weights and scale that seemed to be sticking to the underside of the table. “Oh. Weird! All the writing is reversed.”

“And look. It makes the setup exactly symmetrical from top to bottom,” said Noah. “It’s like a perfect mirror image.”

Dana grabbed one of the reversed weights and put it on the bottom of the antigravity machine’s non-moving belt. Then she looked through the machine from the other side. “Hey! Look at this.” Viewed through the machine, the weight was right side up on the belt and the letters and numbers on it looked normal rather than reversed.

“Wow. What is going on?” asked Noah.

“The machine is warping reality on the way through somehow. But don’t ask me how.” said Dana. “This is really wild!”

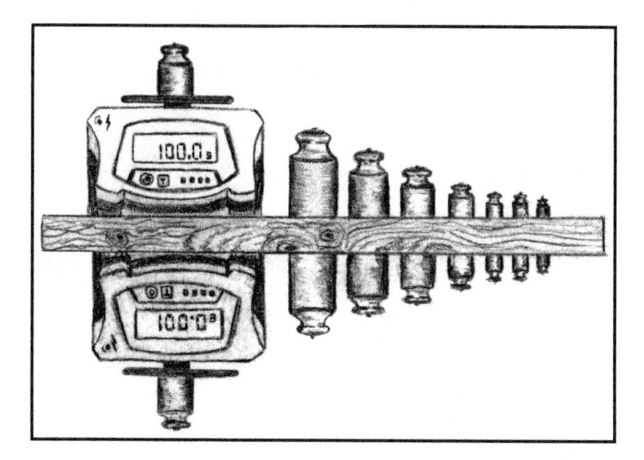

From the notebook of Dana Dash

With their mirrored work station set up, they began experimenting with different combinations of weights, linking them together with a piece of wire. Some of these experiments confirmed exactly what was intuitively expected. But Dana noticed that combined positive and negative weights felt strange in her hands, so she experimented with the larger weights to see if she could amplify the effect she was feeling. With bigger weights, the strange feeling as the weights rested in her hands was even more noticeable. On impulse, she dropped the combined weights and watched them float slowly down to the floor. They picked up speed but not like a normal dropped object.

"Hey Noah! Come check this out." She repeated the drop for Noah's benefit.

"Whoa!" said Noah, "Is that weird or what? I know that connecting the neg-grav weight makes the combined system weigh less, but I thought objects are always supposed to fall at the same speed, no matter how much they weigh, no?"

"Yeah. But I think I figured it out." Dana opened her lab notebook. "You know this equation?" She wrote:

$$F=ma$$

"Sure," said Noah, "Force equals mass times acceleration. Very basic physics."

"Right," said Dana, "But the force of gravity also increases with mass. So mass is really on both sides of the equation. Increase mass, and the equation stays balanced. The acceleration will stay the same. That's why things fall at the same speed, no matter what they weigh."

"Ok," said Noah. "I get it. But what's happening with our combined pos-grav and neg-grav weights?"

"Is that what we're calling them now? Pos-grav and neg-grav?" Dana asked.

"I'm trying it out," said Noah.

"Ok. Let me know how that goes. Anyway, the neg-grav... oh no, you've already got me doing it... ok, the neg-grav weight reduces the total weight of the system but it still increases the inertia,[58] just the way a pos-grav weight would. So, we have the gravity of Earth acting on a combined mass of..." She wrote:

30g – 20g = +10g (gravitational mass)

"But then you have inertial mass:"

30g + 20g = 50g (inertial mass)

"Which makes the acceleration, um... I work it out to be just under two meters per second squared.... Ah! I see. It's exactly one-fifth of the normal Earth gravity. Makes sense, ten grams with fifty grams of inertia. Ten divided by fifty, so it's falling like it's in one-fifth gravity... that's almost like being on the Moon."

"Woah," said Noah. "I bet we can come up with some seriously cool ways to use this effect. Hey! Let's try the same thing but in reverse. Combined some weights so they are slightly negative."

"Good idea," said Dana. "Rather than putting a dent in our ceiling, they should float gently upward, like a helium balloon."

Dana, watching Noah work to combine the weights, stacking them on the neg-grav scale under the table. She was recording the values in her lab book when she noticed something strange.

"Hmm... that's funny..."[59] Dana said.

Bosco, napping under a table on the other side of the room, had ignored their previous talk, but something strange in the tone of Dana's voice caused him to open up one eye and thump his heavy tail against the floor.

"What?" Noah asked.

"There!" She pointed to the scale on the top of the table. The wire

[58] *Inertia* keeps things from moving unless force is applied. Like when you can't get out of bed in the morning until your parents yell at you, that's because you have too much inertia.

[59] "The most exciting phrase to hear in science, the one that heralds new discoveries, isn't 'Eureka!' *(I found it!)* but rather, 'Hmm... that's funny...'" — Isaac Asimov (1920–1992)

used in the previous experiment sat on the test plate.

"The piece of wire?"

"The weight of the wire. It changed!" The scale's digital readout displayed 0.98 grams. "It was exactly 1.00 grams before, remember?"

"A calibration problem?" Noah guessed.

"Maybe," said Dana. But she felt a familiar little jolt of electricity in her spine that meant that she didn't believe it was simply an error. "Take the weights off the neg-grav scale," she said.

Noah removed the weights he had placed on the lower scale. The upper scale display immediately changed back to 1.00 grams.

"Whoa!" said Noah. "Why is it doing that?"

"I have no idea," Dana said, "But this reminded me of something I read about the discovery of infrared light by a brother and sister team of astronomers.[60] They were using a prism to break light into a rainbow of colors and measuring the temperature of those colors with several thermometers. They also had a control thermometer measuring room temperature. But weirdly, the room temperature was reading as hotter than any of the light colors. That couldn't be right. Any amount of sunlight had to be hotter than none.

"So the brother gets another thermometer and finds that the room is indeed cooler. So maybe the other one is broken, right? But then his sister points out to him that she had set the control thermometer exactly where the next band of light would be if there was one. Like, if there was an invisible color before red it would be the hottest color. Infrared light!"

"So they made a big discovery by accident," said Noah.

"Right! Some people say that necessity is the mother of invention, because when people need something they go looking for it. But finding

---

[60] Frederick Wilelm Herschel (1738–1822) and Caroline Lucretia Herschel (1750–1848). Caroline is only credited as Frederick's assistant on most of their discoveries, including hundreds of double star systems and the planet Uranus, as well as discovering infrared light, but she is given solo credit for several comets, a companion galaxy to Andromeda, and a number of nebulae.

something you're not looking for is sometimes even better."

"I guess that makes accident the father of invention," said Noah. "So ok, have we now discovered infra-grav in addition to neg-grav, or something?"

"Something," said Dana, "Definitely something. We are now interfering with gravity. If we can figure out how to do that in a controllable way, we really will be able to fly!"

CHAPTER 13 — LEARNING TO FLY

Bacteria was no more interesting than grass, insofar as watching it grow was concerned. So, while they waited, Dana and Noah continued to experiment with what was happening to interfere with the other scale when neg-grav and pos-grav objects touched each other.

"I think I figured it out!" said Dana. "Look at this. I took two cardboard drink coasters and ran one through the machine, then glued them together."

One side was labeled with a plus sign and the other a minus sign. Dana flipped the cardboard and glue sandwich into the air.

"Call it," she said, "heads or tails. Plus or minus."

Noah called out, "Plus!"

But instead of either falling or rising, the spinning coaster coin did a weird dance in the air. It fluttered, sometimes lunging quickly, this way and that, up and down. Depending on which side faced the ground, it either rose or fell, but its light weight would almost always allow air resistance to flip it over before it got too far in any one direction.

"What is happening?" asked Noah.

"The contact layer between neg-grav and pos-grav objects is a grav-shield. I don't know why it works but that is what is happening. Now, whichever side is pointed down towards the Earth is the one that determines whether the weight is positive or negative. The side away from the Earth becomes gravitationally inactive because the contact layer shields it. The Earth can't *see* that side." She did air-quotes around the word "see."

She grabbed the flipping coaster out of the air and handed it to Noah who set it on a table, minus sign up, where it stayed put, precisely the way a coaster is supposed to do.

"Hey, I wonder..." Pulling a dime out of his pocket, he set it on top of the coaster sandwich. Dana and Noah stared as it floated up above the coaster a few centimeters, then slowly settled again, only to bounce off the coaster and rise once more.

"Cool! there's no gravity at all above the coaster!" said Dana.

"Hmmm," said Noah, "If we could make it bigger, we could have a lot of fun. The machine can't take in anything bigger than maybe fifteen centimeters[61] wide. But... there's no limit to length. We could, say, pass as much rope through that opening as we wanted to."

They looked around the room for some things that were long, but not more than 15 centimeters wide. Noah spied a roll of construction paper which he then passed through the machine. Then, by gluing paper from another roll to it, he created large gravity barrier sheets to lay out on the floor.

"Watch this," he said to Dana as he ran towards the paper covered area on the floor. He jumped, and sailed weightlessly across the paper. He continued to gain some altitude as he passed over the paper and so, landed a bit hard on the other side.

Bosco, excited by the action, got up to play. Unaware of the physics at work, he scampered onto the gravity trampoline to play along, and was very surprised to find himself floating slowly upward over the paper gravity shield. He communicated this surprise by squeal-barking like mad, in case his sudden weightlessness had escaped Dana and Noah's attention. Dana managed to grab his collar before he drifted too high and dragged him back down. He wasn't hurt but, now unsure of the operating rules of ground and air, he looked at the floor under his paws suspiciously for quite a while.

"I wonder how high he could have gone," said Dana. "To the ceiling? Higher? Could you stand in the center of this paper and launch yourself into space?"

Noah said, "Hmmm... I guess the roof over that section has no weight right now. And the air too. And the air above the roof all the way up?"

---

[61] About six inches.

"That can't be right," said Dana. "Air pressure is created by the weight of the air above pushing down. If the air above us had no weight, the air around us would be pushed in from the sides, forcing the weightless air upward. It would blow the roof off and carry it into outer space. Probably cause a mega-hurricane."

She thought for a moment about the potentially serious consequences of their experimentation.

"No," she said. The higher up you go, the more of your weight will come back."

"What?" Noah said. "You think it only works close up?"

"No. Not exactly. Look at this," she said, pulling out paper and pen. "When you do a physics problem, you tend to think of the mass of the planet acting from its center. But that isn't really what is happening. Every little bit of matter attracts every other... or usually does, unless you pass it through Dr. Cavor's machine.

"So look. When you're close to the gravity shield, you're completely in its 'shadow.' But the higher you get, the more of the planet can 'see' you from the sides."

"Ah, I get it!" said Noah. "So I gain weight as I rise and lose it as I fall."

"Right!" said Dana, "So if you stay in the center you won't come down hard. But if you slip off to the side while you are up there, you'll fall to earth like an apple. So be careful!" Dana grinned and leaped into the air above the gravity-shielded zone, sailed all the way to the ceiling, pushed off, and landed perfectly a bit past the far edge of the paper.

"Safety third!" Noah laughed and tried to copy the gravity-vaulting maneuver.

For half an hour, they played like kids on the most unusual trampoline ever. Bosco even got the hang of it, and seemed to be enjoying himself, bounding through the air — right up to the point he scratched at the lab door to go out. When released, he promptly threw up on the lawn and began eating the grass.

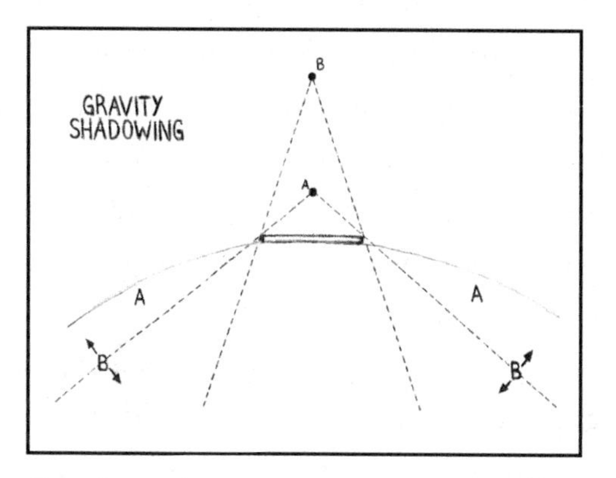

From the notebook of Dana Dash

What finally put an end to the game, as with most really fun things, was someone getting hurt. Dana bumped into Noah midair, bouncing her out of the gravity-shield-fun-zone, landing with a heavy thump on her back at the edge of the paper. For several seconds she lay there gasping for air like a fish out of water. Noah rushed over to her, helped her sit up, and patted her on the back until she regained her breath.

When she stopped wheezing he said, "Boy, you really dashed into the ground. Dash by name — you sure do dash by nature!"

She pushed him hard and he went flying backwards over the fun zone, laughing uncontrollably. He landed on the other side amusingly hard, but entirely unhurt, and lay there laughing on the floor.

The next day, not to be outdone by Noah Knight, widely famed inventor of the neg-grav-trampoline, Dana Dash came up with a new idea. It was not quite as cool as a flying surfboard, but she thought it might work well for personal flying.

It was only a few minutes of work to build a small version of the "flying basket," as Dana had decided to call it. She called Noah over from checking on the tanks to show him her model. It was a ceramic cup, coated with glue inside, that had one of the weights stuck to the bottom. Dana placed one of Bosco's chew toys in the cup, to represent a rider, and gave it a push in the air. It glided across the room like a sailboat on a pond.

"Woah. Seriously cool!" said Noah. "How does it work?"

"The cup is neg-grav, but it is coated on the inside with pos-grav glue. So the whole inside surface becomes a gravity shield. It doesn't matter how much weight you put in it, because there is no gravity in there. Just like above the gravity trampoline."

"Why does it stay at that same level and not go up or down?"

"The pos-grav glue on the bottom that holds that normal pos-grav weight means the Earth can't *see* the bottom of the cup, only the sides," she explained. "The weight acts as a keel to keep the cup upright, and it balances out the lift of the cup, but only at a certain height. Just like with the gravity trampoline, as you go higher or lower, the amount of the Earth that can *see* those neg-grav sides changes. So it has a natural height it will stay at or return to." She pulled the cup down slightly and it proceeded to bob up and down like a boat on water.

Noah was obviously impressed. "If we can figure out how to make one big enough to ride in... I guess we would just need an electric motor with a fan or something to steer it."

Dana replied, "I have an idea for that too. All you need to steer is a pole with a weight on the end to act as a rudder, except instead of going only side to side, you can go up and down too. The Earth's gravity will act as the engine and provide the force. Stick the pole out of the basket on one side and gravity will pull you towards the part of the planet not shielded by the basket. Away you go! Aim the pole at the ground, gravity pulls down on the weight like an anchor, and you can land. If you add a neg-grav weight on the other side of the pole, you can use that side to push off of the Earth's gravity, and go higher. Up or down, forward or back, depending on which end of the pole you hang out, and which end you keep shielded above the basket."

"Well," said Noah, "sounds like all we need to do is build one big enough for both of us to ride in!"

“The machine’s opening isn’t very big. But if we can figure out how to build a bigger flying basket, we really need to build two.” Dana grinned. “Otherwise, how are we going to sky-joust with our weighted-poles?”

Noah laughed. But nervously like he didn’t know if Dana was kidding. “Ooh!” he exclaimed. “I think I know how to build a bigger one. Your calling it a basket gave me the idea. Have you ever seen a rope basket?”

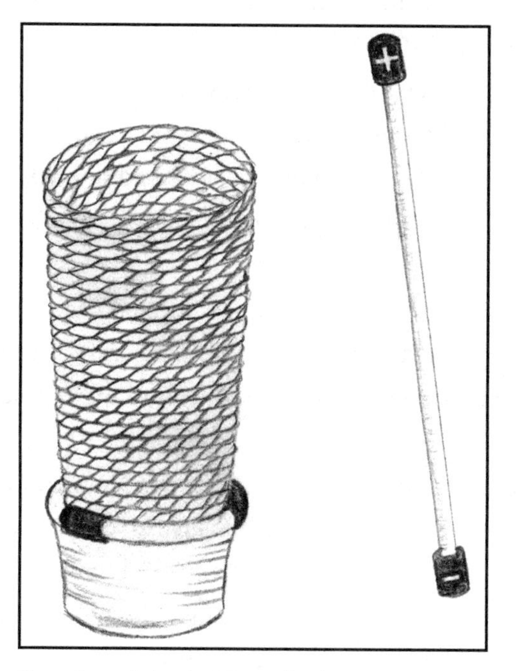

From the notebook of Dana Dash

Neg-grav rope, sharpened steel rods pushed through it to hold the shape, and a quick-hardening resin compound they found on one of the lab shelves did the trick. A keel was created by securing a round tub to the bottom, and by adding or removing sand from the tub, they were able to set the natural flying height.

Much of this work had to be done on the ceiling, at the top of ladders, working with materials that did the exact opposite of the normally expected. Resin, applied too thickly, would drip upwards, for example. They found out that constantly working overhead was much harder than doing the same work at floor level. By the end of the day, they had built only one flying basket, and both children were exhausted.

Noah checked his bacteria tanks and decided they would be ready by morning but they didn't want to end the day without at least a test run of their invention. Dana was climbing into the floating basket, with one foot still on a ladder, when they heard someone enter the office.

"Anyone home?" Prima called out.

Dana tried to reverse direction, but only succeeded in pushing herself off the ladder and thoroughly inside the basket, where she floated above the open door. Prima came through the door and stood directly below her.

Don't look up don't look up don't look up, Dana thought, as she crouched weightless inside the hovering basket.

"Hey Prima! How are you doing?" said Noah.

He quickly walked over to try to keep her from coming any farther into the lab and tried to maintain eye contact to prevent her from looking around.

"Just heading up to and wanted to see if you were done working and wanted to walk up with us," said Prima. "Is Dana around?"

"Uh... no," said Noah. "She isn't here... right now..."

"Oh good!" said Lemma, coming into the room beside Prima. She looked past Noah at the completely assembled Bio-Dodecahedron Lander. "So that's the biosphere you'll be living in for a week? Why does it need legs?"

"We call it the Beetle," said Noah. "It's supposed to look like a spacecraft."

"Beetle, huh," said Lemma. "Dumb-sounding name for a dumb-looking piece of junk. But you're not dumb. For a boy, you're pretty smart. Too smart to waste your time in Tech. If you go to school here next year, you should consider joining Phi. You'd be better off if you dumped that dumb Dana Dash and found a competent lab partner to do some real science with."

Dana seethed with anger in the basket above Prima and Lemma. She wanted to drop the heavy end of her weighted pole on Lemma's stupid head but she controlled herself somehow. She didn't know exactly what would happen if the science kids discovered their gravity-reversing secret, but she knew Dr. Cavor would find out Dana had taken her machine. Lemma was a snitch for sure.

"Be nice," said Prima.

Lemma snorted, and tried to walk towards the Beetle for a better look. Noah stepped into her way and said, "I thought you were going to eat."

"Hey," said Lemma, craning her neck to see around him. "You built that pretty big. Don't you need to have it up at the gymnasium where the contest finals are? The judges probably aren't going to come down here. How do you plan on moving it? Will it even fit through the door? Doesn't look like it will."

She snickered as she turned and walked back through the office door.

"Are you coming?" asked Prima, starting to follow behind Lemma.

"I'll catch up," said Noah and waited until the science kids were out of earshot before looking up at Dana in the hovering basket.

Dana peered over the edge of the basket and saw the expression on his face. "Well go on. Don't just stand there. Get a tape measure and let's find out how tall that roll-up door is."

"I hate to admit it, but Lemma was right," said Dana, after she had done some measuring. "It doesn't fit through the door, we can't disassemble it because of the welding, and all the tanks are mounted in place and already full." Then she grinned widely.

"Yeah, doesn't sound promising," said Noah. "So why are you smiling?"

"Ok. Look at how big the skylight section of the roof is that slides back. It's bigger than the roll-up garage door. So — I'm thinking we go out the sky-door instead."

"Since we don't have a large crane handy, you must have some neg-grav trick in mind."

"Yep. The aluminum frame is strong but actually pretty light. The window plastic weighs more than the frame does. But, as I calculate it, the water in those tanks weighs more than all the rest of it combined."

"Ok," said Noah, "So if we convert all the water in the tanks to neg-grav, we can float the whole thing up through the retractable sunroof. Then how do we stop it? We don't want it flying into space."

"That's the cool part. I already installed computer-controlled shutters on the outside of both the Beetle and Beetle-heart, so we could both have privacy and control the UV[62] light levels in the bacteria tanks. So all I have to do now is add a layer of neg-mass to all the shutters to make them into gravity shields..."

"... and the Beetle becomes something like the flying basket," Noah said, nodding.

"Better than the basket. Way more control. The shutters give us total control of push or pull. The landing plate acts as the keel. And, since the neg-grav push is on the core, not from the sides like it is in the basket, exposure isn't reduced as it goes higher. We wouldn't even have an upper altitude limit. No problem getting it up and out. We land in front of the gymnasium and Lemma will go crazy trying to figure out how we moved it!"

Noah smiled a slow smile. "Let's do it. What do you need from me?"

"First we need to eat. I'm starving. And I've been grabbing something for Jayla too — she said she's too close to being done and doesn't want to go all the way up the hill for dinner. Can you get us some grub while I hocus-pocus these shutters into gravity shields?"

"Blech," said Noah. "I hope it's better than grub today. You never know what they'll have laid out for us. I'm tired of spaghetti for breakfast. But yeah, I'll hit the weirdo-daily-sustenance-pile and bring back enough for the three of us."

---

[62] *Ultraviolet light* (UV) is invisible, like *infrared* (IR), but on the other end of the spectrum, way past blue. It is good for making plants grow and giving you a sunburn. It's what your sunscreen blocks. Some inks *fluoresce* (produce light) in UV light by converting it to visible light. Great for secret writing!

When Noah got back, Dana had already made a good start on the neg-grav retrofits. She looked skeptically at the array of food-like items he laid out on the lab table.

"What are those?" she asked, pointing at an unappetizing pile of bar-shapes.

"They had a bunch of protein bars I thought I'd stash in the Beetle. And I dropped off some food for Jayla, and made apologies for you not being able to come. She knows since we're camping in the Beetle for a week that we need to be done a lot sooner than everyone else, and she says she understands if you can't come play radio ham."

"I'll still try to stop by and see her later. She said she'd look after Bosco while we're in the Beetle, and I'll need to drop off his food. But I figure Chip can stay with us; he doesn't breathe much and his food doesn't take up much space. We'll have to clean out his cage a few times, is all."

"Oh, good thinking. But," Noah hesitated, "can Jayla handle Bosco? I mean..."

"Jayla is certainly capable of handling Bosco. Don't underestimate her just because she's in a wheelchair. She's probably the most capable person I know. Except for you, of course, my partner in crime." Dana grinned at Noah.

Noah rolled his eyes in response. "Yeah yeah yeah, I know, you couldn't do this without me. So what else can I do now that I'm done with waiter duty?"

"We need a way to run the water in the tanks through the gravity-reversing machine, so that it converts to neg-grav. Can you do that?"

Noah thought for a minute. "How about running a hose through the machine? We can drain the tanks through one valve while refilling them through another. Gravity changing direction halfway though should make it all work like a circular siphon. It'll be like it's always running downhill, even though the water ends up back in the same exact tank."

"I think that should work," Dana agreed. "Okay, see if you can find a hose; I'll keep working on the shutters. That much water has quite a bit of mass — we'll need the inner grav-shields in place before we do this. If we don't keep it grav-shadowed, the push from the Earth could send the Beetle crashing through the ceiling and into outer space."

By the time Noah returned with a long garden hose, Dana had finished coating the insides of all the shutters with a layer of neg-mass resin.

"Where did you find that?" Dana asked as he flopped it next to the machine.

"Garden at the manor house. Was next to that sad-looking scarecrow. It just looks like a sad bag of clothes without the hat. Don't worry, I'm just Robin-Hooding the hose. I'll put it back." He grinned at Dana and she stuck her tongue out at him, which made him grin even wider.

Dana connected one end to a tank valve carefully while Noah threaded the other end of the hose through the machine and back to the second water tank valve. However, the second connection didn't work right.

"I can't get this threaded on the spigot," said Noah.

"Hmmm," said Dana. "Oh, look at this! The thread goes the wrong way."

"Oh whoa!" Noah laughed. "Just like the writing on the weights. It got reversed inside the machine."

"Hold on, I've a compression coupler somewhere here." Dana found a short rubber cylinder, with metal ratchet ties on either end, that fit over the hose and spigot. Tightening the ratchets with a screwdriver, she made the seal watertight. "Ok, that should hold. Let's try it."

They opened the valves, and a steady stream of the faintly glowing bacteria water began to flow. The fluid was thick with bacteria and flowed smoothly but slowly, like gravity-defying maple syrup poured on a stack of pancakes.

"It's working!" said Noah. "Look! The returning water is pooling on the top of the tank while the bottom water drains away."

It was true. The small air gap that had been at the top of the tank was now working its way down to the bottom — an indicator of how much of the tank's contents had been converted to neg-grav.

"Genius," said Dana, smiling appreciatively.

"Aw shucks," said Noah.

"You are though! That water is so bluish-green colored it looks like some sort of cyanobacteria sports drink. Cyan-aid! Because it's cyan[63] colored and aids our breathing!"

"Ha. Sounds a bit like cyanide... actually many cyanobacteria strains do produce deadly poisons called cyanotoxins. But I made sure we are using a completely safe kind. Not that I would recommend drinking it!"

Dana laughed. "Noted. Don't drink the cyan-aid! So, is it ready?"

"Sure is. Everything's ready to go on Friday, as far as I can see."

Dana nodded. "Just gotta get this baby over to the gymnasium."

She disconnected the tank, and they went through the same process for each of the other water tanks, including the tanks they had set aside for drinking water, until all the liquid at the center of their biosphere project had been run through the strange machine and would give them the neg-grav they needed.

The Beetle was ready to fly.

---

[63] *Cyan* is a blue-green color, and where the cyano-bacteria gets its name. Cyan is also the color of the planet Uranus, which gets its color from clouds of methane gas. Yes, go ahead and laugh that Uranus is full of methane. If you don't get the joke, ask a grownup. We can't print it here.

CHAPTER 14 — ESPIONAGE

Dana nudged Noah. "Now we just need to wait until dark and we can pilot this thing over. Won't everyone be surprised when they see our project sitting outside the gym!" she said excitedly.

Noah wasn't listening to her. He was looking around, scanning the area for something. He stepped through the door into the office section of the lab and looked there too.

When he came back through the door, Dana said, "What is it? What are you looking for?"

"Where is your dog?" asked Noah.

Dana looked around too. Slowly at first, expecting that Noah had simply missed Bosco laying down under a table somewhere. Together they scoured the whole lab, but Bosco was nowhere to be found.

"The doors were all closed. How could he be gone?" She wondered out loud.

"Maybe when I went to get the hose? I don't remember the last time I saw him," said Noah, tugging on his shirt tail.

Dana grabbed the leash and they hurried outside. "Bosco!" Dana yelled, "Here boy! Here boy! Treat treat treat!"

"Let's split up," said Noah. "I'll look over by the manor house. Maybe he followed me up there."

"Ok," said Dana. "Meet back here in fifteen minutes, whether you find him or not. I hope he didn't try to go home..."

Dana headed up towards the cafeteria, half expecting to find him waiting in front of the door, ready to eat. He wasn't there. She walked over to the dorms, hoping he had gone there. No Bosco. Dana was at a loss for where to look next. She wandered around the school grounds between the empty classroom buildings, carrying a leash and calling Bosco's name, her worry increasing with each Bosco-less minute that passed.

After about ten minutes, someone called out to her.

“Hello child! Is this the Bosco you seek?”

Dana turned to see the giant woman who had almost caught Noah and Dana in the abandoned preschool. She was striding towards Dana, looking much the same as when Dana last saw her: twisted nest of blonde braids around her head, overalls with a patchwork of pockets, and a plaid shirt, though this time it was a cheery pink instead of red. On her face was the same large eyepatch covering her right eye. And there, looking like a puppy cradled in the crook of one of her huge arms, was Bosco.

Dana’s relief at seeing Bosco safe was almost eclipsed by her marvel at the woman’s massive body. She was as proportional as a healthy, regular-size woman, but almost half again as large as any other person Dana had ever seen — half again as large in every dimension. Dana did some quick calculations in her head: *three dimensions each 150% size is* $1.5 \times 1.5 \times 1.5 = 3\frac{1}{3}$ *times the weight of an average woman. She might weigh well over two hundred kilograms!*[64] She wanted to ask, for the sake of scientific curiosity, but knew that would be very impolite.

Dana looked back to her dog, who was panting contentedly and seemed to be enjoying riding around on a human, which he hadn’t done since he was a puppy. “Oh thank goodness,” said Dana. “I was worried he tried to go home. He’s staying with me while my parents are on vacation. I’m Dana Dash. I’m in the STEM contest.”

“Hello, young Dana. I am Master Skadi. I am the keeper of security here at the Cavor School... among many things. Your dog found his way into one of the school buildings. I discovered him in Dr. Taxidiótis’ office, on her desk. I hope he caused no damage to her computer. He was pawing upon her keyboard.”

“Bad dog!” said Dana, laughing. “Stay off other people’s computers. No espionage!”

[64] Maybe 450 lbs.

The woman bent down and gently placed Bosco on the ground next to Dana. He laid down and panted up at Master Skadi appreciatively. The woman then straightened up to her full height in front of Dana. She stood there for several moments eyeing Dana appraisingly, and then slowly began to walk in a circle around behind her and Bosco. Dana turned, trying to keep the woman in front of her, wondering what she was doing.

Master Skadi finally stopped turning and put her hands on her hips, like a defiant warrior. "It makes sense to me that this trespassing dog is belonging to you."

"What... why?" asked Dana.

"On a day in the past, you and your companion were in the childcare building where you should not be. On this day now, you are wearing a picture of Einstein with a front serious-face, and back not-serious-face. On the back, the tongue of Einstein is sticking out. It is the same shirt you wore on that other day."

"Oh," said Dana, feeling her face flush with embarrassment. "I guess that is distinctive."

"Indeed," said Master Skadi, "It is unlikely that it was another small girl-child who ran away from me with Einstein mocking me from her back."

"I'm sorry," said Dana. Her heart plummeted as she realized that this woman could have her thrown out of the contest, could crush her dreams of going to school here. Dana stumbled over her words to explain. "I wasn't stealing or breaking in — it was my preschool, my dad told me — I wanted to see — to see what I could remember." She gulped. "Please please don't say anything... I want to go to school here more than anything!"

The woman towered over Dana, staring down at her with one frosty blue eye. Dana bent to connect Bosco's leash, but also to avoid the woman's gaze. Tears were welling up in her eyes, but she wasn't going to let the woman see her cry.

"Peace, child," said Master Skadi, bending down to Dana's level, "I will not tell, of this dog's trespass, nor of yours and your companion's. It is unlikely that it would be of concern to the others. There are such things here that..." She stopped, and her face softened a little. "Go, back to your contest, youngling."

Dana felt a surge of relief. She blinked away the tears, that both she and Master Skadi were pretending weren't there. "Oh, thank you — thank you," she said as she gathered up Bosco's leash and turned to go. Then she remembered something.

"Hey um... your knuckle tattoos..."

"Ah, yes. They are nothing." She threw back her head and laughed loudly at her own joke.[65]

When she stopped, Dana said, "I recognize Euler's identity on your right hand, but what is the other one?"

"You are indeed a smart one for your age," said Master Skadi, thrusting her fists towards Dana as if she were throwing a double punch. "My left hand is Dirac's equation.[66] It is unifying in physics, on the same level that Euler's Identity is unifying in pure mathematics."

"That's super cool," said Dana. "I'm going to look it up."

"I believe you will. You have purpose, young one, I can see that in you. But first you must focus on the task at hand, to prevail in your challenge with the other Disciplines. I am a Technologist also."

She lifted her eyepatch slightly and scratched underneath it with a large finger. "Go. Take your Bosco, and make our Discipline proud."

Dana held tight to Bosco's leash, just in case he got any more running-away-to-break-into-offices ideas, and hurried to meet Noah, who by now would be waiting and probably worried. Noah, waiting exactly where he said he would be, tugging at the hem of his shirt, was visibly relieved to see Bosco.

---

[65] Both of the formulas on the knuckles of each hand evaluate to zero. Hence, "they're nothing."

[66] Paul Dirac (1902-1982) formulated the first equation to tie together the theory of special relativity with the principles of quantum mechanics. In natural units, the *Dirac equation* is expressed as $(i\not{\partial}-m)\,\psi = 0$.

As they walked down the hill towards the lab, they saw a girl walking out of the open door of their lab, which they had left ajar, in case Bosco returned like the good boy he wasn't. The girl walked out of their lab, and into the next door — Lemma and Prima's lab. "Hey!" exclaimed Dana. "Did you see that? Someone was in our lab!"

They hurried back to the lab and searched for any signs of tampering. Dana had flashbacks of sitting in Principal Peters's office, and of Matt McCaws's smug face as he won first place in the science fair. She hoped she wasn't being paranoid — but in a few days they would start their week in the biosphere, so this would be a particularly bad time for sabotage.

Dana tied Bosco to one of the Beetle's support legs, so he couldn't run away again, and looked over the structure carefully. There was nothing visibly damaged. They'd only been gone fifteen minutes chasing the dog, so whoever it was couldn't have been in their lab very long. Maybe the girl hadn't done anything. But what if she had? What if she was right on the other side of that wall, gloating about how she had ruined Dana and Noah's project?

"Noah," Dana called out to him. "Check the cyan-aid and make sure it hasn't been turned into cyanide or whatever? I'm going to go see if I can listen in on our next-door neighbors. If they did something, I'm sure they'll be talking about it."

"Checking them now," said Noah. "But how are you going to listen in?"

"Hmmm," said Dana. She looked around, and her eyes fell on the floating basket, tethered at the top of the stepladder.

"Open the sky-door," she said. "I have an idea."

As Noah activated the roof, Dana climbed up the ladder and into the basket. It was a very odd feeling, going over the rim of the basket into the weightlessness inside. She wasn't quite used to it yet. When Noah had the roof open all the way, she braced her feet against opposite sides of the basket, briefly wishing they had thought to build in loops to hook her toes under as she pulled out the double-ended weight-pole and pointed the neg-grav side towards the Earth. The basket cruised upward at an angle through the open roof as if she had been steering a gravity-reversing-sky-basket her whole life.

Dana quickly traveled far enough to be over the Science kids' lab. She pulled the steering pole back inside to slow down. When the basket settled to the roof, she climbed out as carefully and as quietly as she could. She slowly moved over to their retractable roof section, which they had left slightly open, probably just to get some fresh air. Through the crack, she could see three figures below her in the Phi kids' laboratory area, and she could hear them talking quite clearly through the glass. Up close, the mysterious lab trespasser looked like a carbon copy of Lemma, only slightly older and taller. She had to be the sister Lemma mentioned the first day, the one who thought it was so important to sit at the Phi cafeteria table.

"Why the big cage?" asked the sister. Dana saw what she was talking about. A huge wire mesh cage took up about a third of the lab space. It was big enough to contain several lab tables full of equipment inside it.

"We were getting too much interference on the bioactive frequencies," answered Lemma. "I asked Mother about it. She said the school was deliberately jamming those frequencies but when I asked why she said she couldn't tell me, but that our project was definitely on the right track."

"Anyway," said Prima, "we had to build a Faraday cage.[67] We even got a bunch of electrically conductive fabric to lay over the lab table and on the floor. We'll be completely isolated from any radio frequencies."

"Well," said the sister. "It looks impressive. I don't think any of the other contestants have anything that can top this."

"Great," said Lemma. "So, Lambda, can't you just let us win this on our own? You don't need to do anything."

---

[67] Michael Faraday (1791-1867) discovered in 1836 that a tight wire mesh was enough to effectively block all electromagnetic frequencies, isolating the inside of the "Faraday cage" from outside radio signals.

"Sorry. Even if you don't win, Dana has to lose. The only way she can afford to go to this school is with that scholarship. Mother doesn't want that to happen, and she's not willing to leave it to chance."

Mother? Why would Lemma's mom care if Dana went to the Cavor School? Maybe Lemma would be happier, but then her sister Lambda wouldn't be saying to do it for their mother.

Dana missed what was said next while she was processing, but caught the end of it. "...no, she wasn't there," said Lambda. "I didn't have much time. But I have an idea..."

"I don't want to know about any of this," said Prima. "And whatever it is, I don't think you should do it."

"Yeah," said Lemma. "Actually I feel the same way. I don't particularly like Dana, but why does Mother care so much if she goes to school here?"

"I don't know. But I'm not going to tell her no. Are you?" Lambda paused, and when no one said anything, she continued. "That's what I thought. I'm going home now to let Mother know what's going on. Keep up the good work."

Dana was seething — a hot, almost palpable anger. She thought about Matt McCaws sabotaging her science fair project at Hillhurst School — now this! It wasn't fair. Her fists had clenched at her side; she actually wanted to hurt Lambda. Or maybe it was their mother she wanted to hurt. Through the roof she watched Lambda head for the lab's exit, and quickly scrambled back into the basket. She wasn't sure what she was going to do, but Lambda wasn't going to get away with this.

She lifted off the roof and watched as Lambda left the lab building. The sky was getting dark and the lamp posts that illuminated the school walkways had just turned on. Dana cruised above them, making sure the flying basket would be hard to see above the lamplight as she followed silently about five meters[68] above Lambda's head.

---

[68] This is not very high. Look up above you — would you notice somebody flying over your head?

She wrestled with herself; what she should — or could — do. She pointed the pos-grav end of the pole at Lambda, and the basket was drawn downward towards the sneaky saboteur. *Going to ruin my project, are you? Stop me from going to this school, will you? Certainly be a lot harder to do if you had a big dent in your stupid head...*

As she silently sped closer to the unsuspecting girl, the voice of her karate instructor came into her head. *We must never be the first to resort to violence.* She sighed. Violence was not in her nature. Or at least it shouldn't be. Her amygdalae[69] and adrenal glands[70] were telling her something very different right now. They didn't want to hear about her better nature.

With another sigh, she reversed direction on the steering pole, and felt the Earth push against the neg-grav as she silently floated backwards into the dark sky away from the entirely unsuspecting Lambda Lei.

But as she watched below her, something unexpected happened. Lambda was crossing one of the illuminated circles of lamplight when something appeared out of the darkness, moving fast. At first, Dana's mind saw a sort of giant brown centipede and she almost cried out a warning. It was a train of four deer, linked and moving together like a long multi-legged arthropod — *a deeripede*, her brain suggested not at all helpfully — each holding the tail of the deer in front in its mouth. It was such a confusing sight that the warning yell died in Dana's throat.

The thing was upon Lambda before she even noticed it. The front deer was a buck, with something tangled in its antlers — almost like a second head made of rags and plastic bags. It looked like the one in the bushes by her house weeks ago, but its antlers were even more piled with garbage now. *How many trash-head stags are running around Dash-on-Hudson,*

---

[69] The *amygdalae* are small deep brain clusters, sometimes referred to as part of the "lizard brain." They produce quick emotional reactions, fear and anger, that sometimes override rational behavior.

[70] The *adrenal glands,* located on the kidneys, release adrenaline to increase metabolism rates in "fight or flight" situations. This can cause a bad coppery taste in the mouth. Thus the phrase, "spitting mad."

anyway? On top was a floppy, tattered hat... *wait, is that the garden scarecrow's missing hat?* Beneath that hat, a single eye glowed green in the lamplight.

From the notebook of Dana Dash

The stag lifted one front leg high and rested it on Lambda's shoulder. The girl and the entire deer train froze in that position as if ordered to halt. Lambda didn't cry out, or struggle. There was no motion or sound, like a bizarre sculpture.

Dana watched, baffled, as absolutely nothing happened for several seconds. Then the tangled mass of trash moved off of the buck's head and slid onto the motionless girl's back. The scarecrow's hat came with it, like the shell of a hermit crab, as it worked its way around the front of Lamba's frozen body. What was it doing? Why wasn't Lambda moving? Was it hurting her? Dana eased the flying basket down, but still staying out of view above the street light. It looked like the scarecrow hat was rifling through Lambda's pockets.

Dana heard footsteps coming down the path from the school buildings. She looked up and saw a figure walking towards them. The trash-hat-scarecrow-pickpocket-thing hanging on Lambda's clothes heard it too. Tiny pale hands with bony fingers emerged from the tangle of hat and garbage and clawed hastily back up the frozen girl's body and onto the motionless deeripede-buck. For an instant, Dana thought she saw the single green eye glowing in the lamp light. Then the thing was back in its position on the buck's head, between the antlers, and the buck released its hold on Lambda's shoulder. Dana watched as the thing rode away into the night, little claws steering the deeripede by the antlers like a glow-eyed scare-crow-hatted trash-pirate.

Lambda stood completely still for several more seconds, then seemed to jerk awake and look around, startled. The figure coming down the hill moved into the glow of the lamplight — it was Dr. Cavor. She greeted Lambda as she approached.

"Why, hello there, Lambda. Are you here visiting your sister?"

Lambda seemed a bit confused and took several seconds to reply. "Yes. Uh... yes. I talked to her. She and Prima are working late on their project."

"Very good. I am sure their entry will be a credit to Phi Discipline. I am going down to check on my Deltas. Dana and Noah are also working late, I presume."

Hearing her name, and realizing where Dr. Cavor was going, Dana propelled herself back towards the roof of the lab building. She floated the flying basket through the lab's open sky-door and down to the stepladder. She tumbled gracelessly out and quickly tied the steering pole to the bottom. The sky-basket settled gracefully to the floor.

"Noah!" she called out. "Remind me to tell you about some more really weird stuff I just saw," she said, grabbing the gravity-reverser. "But right now we need to hide the weird stuff that's in here. Dr. Cavor is coming to check on us."

"Oh!" said Noah. "What should I..."

"The basket!" she said. Noah quickly grabbed a tarp and threw it over the flying basket, now firmly grounded next to the stepladder.

Dana scrambled into the Beetle with the gravity-reverser under her arm and hastily stashed it between two tanks inside the Beetle-heart, quickly zip-tying it to the aluminum struts. She slid back out of the Beetle just as the office door opened and Dr. Cavor called out. "Dana? Noah? Are you here?"

"Back here working!" yelled Dana. "Come on in!"

"I thought I would see how you two were progressing," Dr. Cavor said, coming through the door into the lab. "Are you prepared to go in a few days? You are planning to be inside your biosphere the entire week before the competition judging, correct?"

"Yeah," said Dana. "We're pretty much done. We're going to spend the next two days doing system checks and then start camping out on Friday morning."

"Splendid," Dr. Cavor said. "Is there anything else you will need of me before then?"

"Um..." Dana glanced nervously at the tarp-covered basket behind Dr. Cavor. It was swaying slightly under the cloth and looked a little like a child's ghost costume on Halloween. "I think we're all set. Uh, Noah, what do you think?"

Noah took the cue to distract Dr. Cavor from looking around the lab and walked around the other side of Dana so Dr. Cavor's sightline was in front at the Beetle, not behind at the ghost-basket.

"Yeah I think we're all ready!" he said. "I put the last of our food rations in today — shelf-stable and high-calorie, like the astronauts eat. Can't think of any other problems. Although you never know — I'm reading a book about the Biosphere 2 project, *Life Under Glass,* by Abigail Alling. Their two-year time frame was much longer, and their biosphere much larger, but they had a lot of problems keeping their oxygen levels up."

"Uh oh!" said Dana, walking towards the Beetle and hoping Dr. Cavor would follow. "Should we maybe read a book by that magician who seals himself up for days inside a glass box? If we could learn how he stays alive without breathing, this will be easy!"

Dr. Cavor laughed. "I am certain you will be able to do this. Would you permit me a tour of your new house before you seal yourself in?"

"Certainly," said Dana. "It's a little cramped, even for us, but you can probably fit."

The children showed their advisor how to climb into the Beetle through the airlock. Bosco, still unhappily tethered to a leg strut, regarded them reproachfully as all three crawled around him and into the hatch. Dana explained the two deck levels, pointing at the hammocks and storage below and the work area above, all arranged around the Beetle-heart dodecahedron with its full aquariums and tubes for water and gasses. Dana also pointed to the aquarium that held the injured red fox. "And this is Chip. He's the Beetle's mascot and will be staying with us."

"An excellent idea," said Dr. Cavor. "Having a much smaller animal with you to be a canary in the coal mine. If anything goes wrong with your air, the fox will be affected first."

"Oh," said Dana, pretending she had thought of that, "Um yes. Indeed. But it should be fine. We have sensors to monitor the air and I've set up alarms if O_2 levels get too low."

Dana pointed out the two laptops in the work area at the very top of the Beetle. "These are the Beetle's brains, where everything is controlled. The left brain monitors temperature and gas levels inside and outside the tanks, and the right brain controls all the shutters. I can open or close the outer shutters for privacy and light. Also the inner shutters around the airlock so we can use that area for bathroom stuff. And the Beetle-heart also has shutters so we can run UV lights at night for the bacteria tanks while we want lights out for sleeping. They can all be operated from this touch screen."

"Very nice work," said Dr. Cavor. "May I have a demonstration?"

Dana thought rapidly. *Which shutters can I open safely?* Currently all the outer shutters were open and all the inner shutters around the tanks were closed. That was how she had set up their default states and how they would stay without activating any of the computer-controlled servo motors. She looked up through the transparent roof of the Beetle. The lab's roof was still open, and she could see the stars starting to come out above her. *Ok, I can demonstrate the shutter controls — I just can't open the ones on the bottom half of the Beetle-heart. As long as the Earth can't see the neg-grav tanks, it can't push on them.*

"Sure," said Dana, reaching over to the power switch. "Give me a minute to fire up the brains and I can show you how the shutters work. It won't be a problem." She knew these systems worked. They had been tested.

What could go wrong?

CHAPTER 15 — SABOTAGE

When the shutter system failed, it failed in pretty much the worst way possible. One second Dana was showing the program on the Beetle's brains to Dr. Cavor while Noah checked the cyan-aid tank levels. The next there was a bump and a sudden plunging feeling of weightlessness in Dana's stomach as gravity-shielding shutters on the outer surface of the dodecahedron and the airlock snapped closed. Then the shutters on the inner Beetle-heart flipped open, and the neg-grav water was exposed to the Earth.

Dana instinctively grabbed ahold of interior support struts at the sudden feeling of weightlessness. Then, as the Earth pushed against the Beetle's central neg-grav and the habitat began to accelerate rapidly upward, some feeling of weight returned. But she didn't release her grip on the aluminum tubing. If anything, she gripped it harder, panic growing from the realization that they were now launching skyward, totally uncontrolled. She let out a groan and heard Noah making a similar sound, no doubt having come to the same conclusion. Somehow Dr. Cavor seemed neither much surprised nor distressed by the sudden strange turn of events.

"It appears you two may have been borrowing some equipment from my personal laboratory," she commented dryly — and unless Dana was mistaken, with no trace of surprise.

Dana didn't answer her. She needed to figure out what had gone wrong with her control program. The software seemed to be running normally, and she wasn't seeing any error conditions, but none of the control movements or commands she tried were having any effect on the situation.

"What happened!" yelled Noah.

"I don't know," Dana called back. "I'm running some diagnostics now." Her ears had begun to hurt and she swallowed and worked her jaw trying to equalize the pressure. "Hey, get that airlock door closed before we get too high to breathe!"

Just then they heard a familiar excited squeal-barking, repeated closer and closer together until they became a frightened howl.

"Bosco!" Noah said and quickly scrambled to the bottom of the flying polyhedron. Looking over the edge of the opening in the bottom of the structure, he said, "He's on the base plate! We need to get him inside with us before he falls!"

Dr. Cavor crouched down and scooted over to the opening Noah was looking through. "Noah, you will need to climb down there and boost him up. I will pull from this end."

Noah climbed down onto the pentagonal base plate. Between the weight of him and Bosco, and under the added acceleration of their ascent, the joining points to the five support struts creaked and groaned ominously.

He put his arms around Bosco and hoisted him upwards. Dr. Cavor grabbed Bosco's collar and pulled as Noah pushed and Bosco scrambled into the habitat. Then Noah quickly followed the dog inside and closed the inner airlock door quickly behind him.

"We have Bosco. What else can we do?" he asked.

"Give me a couple of minutes," said Dana. "I need to reboot the system." Those couple of minutes and many more went by, as Dana tried desperately to figure out why her control program wasn't working. They continued to rise faster and faster, as air resistance outside decreased with greater altitude. The Beetle creaked ominously as the pressure outside lessened and the stresses on the craft's frame increased.

"Nearly in full vacuum now," said Dr. Cavor. "I hope this little spacecraft of yours will hold up to the internal pressure."

Dana was shocked by Cavor's casual use of the word *spacecraft*. She had only set up neg-grav systems on the Beetle so they could move it out of the lab and up the hill to the gymnasium for the STEM contest. At the time, she had known that the system she built had no upper height limit, but hadn't really given the implications of that much thought. Now, hearing Cavor say it, it really hit home that the Beetle — the mock crew capsule they had designed to test their life support systems — was indeed a real spaceship. They were in outer space!

"Don't worry, Dr. Cavor," said Noah, "The good news is, we test-pressurized the habitat on the ground to two atmospheres. It will be fine even if we rise far enough for total vacuum outside."

Dana said, "But the bad news is, I'm looking down through the tanks into the airlock right now, and I can see that you left the outer hatch open. And I'm remembering that we never actually pressurized with the outer airlock door open. I'm pretty sure that changes all the stress patterns — and not in a good way."

"Um..." Noah tried to find a rebuttal to this dangerous oversight.

Dr. Cavor cut in, "I have the utmost confidence in your design and fabrication skills. Despite the serious problems your control systems are obviously now experiencing."

Dana said, "We didn't get a chance to test these controls at all. At least, not off the ground. But I can't figure out any reason for the problem we're having."

Noah asked, "Couldn't we just power everything down and try to operate the shutters by hand? We might be able to get a gentle descent going."

Dana replied, "That is an excellent idea that I totally wish one of us had come up with ten minutes ago. The most important shutters to close are the ones located inside the airlock right at the top. We can't get to them with the airlock evacuated. First we have to figure out how to close that outer airlock door."

"Arrgh," arrghed Noah, "This is all starting to feel pretty doomy." He stroked Bosco, who was still cowering at the bottom of the habitat.

Dr. Cavor said, "I am certain that if we stay calm and focused, we can work this out. Dana, please describe to me how these shutter control systems work."

“Ok, it’s pretty simple. The outer shutters and the ones on the airlock walls control whether or not our normal mass interacts with outside gravity sources. The inner shutters on the Beetle-heart do the same thing with for the neg-grav mass of the water tanks. The brain talks to a bank of wireless relays that activate servos that open or close the various shutters. But see here — the brain says they’re all closed right now, and clearly they aren’t.” Dana handed the computer over to Dr. Cavor.

“I see,” said Dr. Cavor. “Is there any reason why the inner shutters on the water tank would all be open while the outside ones would be closed?”

Dana said, “I did install the servos and shutters differently on the inside. The inner servos open the shutters against their springs, while the outer ones close them. I was trying to see which way worked best. Also the default state when powered down was what I wanted — ” Dana stopped short. “Powered *down*...” She reached over and flipped the switch to cut the power from the batteries to the servo motors.

All the outer and airlock wall shutters opened and all the inner shutters around the tanks closed. Dana felt her stomach rise as the Beetle stopped accelerating.

Dr. Cavor reached over and flipped the switch back on. “Very good,” She said. “Now we know the problem is not in your control program, but in the wiring. Let us keep flying until we locate the trouble.”

It didn’t take long for Dana to trace the problem in the wiring. She opened up the 24-port servo control box she had built and saw the problem immediately.

“Hey! Someone messed with this! This wire wasn’t here. It bypasses the control switches and connects all the servos directly to the power! This is sabotage! I need to remove this,” Dana opened a drawer with some tools in it.

“Wait!” said Dr. Cavor, who was holding the laptop with the control program open. “Do not touch it yet. Allow me to work through some numbers.”

"Shouldn't we stop this launch into space as soon as possible?" said Noah. Bosco squeal-barked and seemed to strongly agree.

"It may be too late for that." Dr. Cavor typed rapidly on the brain. A few minutes later, she said, "In fact, it is too late. We would end up in an orbit."

"What do you mean?" asked Noah and Dana as one.

"Patience. Give me a moment to calculate..." The two children waited anxiously while she worked. "There. Dana, please remove that wire sabot[71] from the machinery."

Dana pried up the staples holding the offending wire and disconnected it from the power. The shutters on all of the pentagons on the top of their spacecraft opened. On the children's side they could look out to see the full Moon.

"Did you do that?" Dana asked Dr. Cavor. "Are we under control again? Can we land?"

"Yes and no," Dr. Cavor said. "I have us under control. But we will not be landing on Earth anytime soon."

Bosco whined again and Dana rubbed his tummy until he began to relax. Dana asked Dr. Cavor, "Why can't we just partially shutter off the neg-grav until we float gently back down to Earth? I don't understand."

Dr. Cavor explained, "Once we were past a certain point, and we had enough side velocity, the Earth's gravity would no longer return us to the ground. Instead, we would miss the Earth and be in orbit. Neither pulling or pushing against the Earth would allow us to decrease that angular velocity."

"I don't understand," said Dana. "Where did the side velocity come from?"

---

[71] As any good linguist, fashion historian, or fan of *Star Trek* movies can tell you, a *sabot* is the wooden shoe from which the word "sabotage" was coined, because cranky people apparently sometimes threw their shoes into machinery. However, since Dr. Cavor was being shot off the Earth into space, the other meaning of *sabot,* a device used to position a bullet for firing, may have been the intended use.

"We had it already when we started," said Dr. Cavor. Producing her own notebook from somewhere, she pulled out a pen and paper and began drawing a diagram. "Standing on the Earth at the equator, we would be moving around the circumference of a forty-thousand-kilometer circle every twenty-four hours. That is over sixteen hundred kilometers per hour — less at the latitude of New York. More precisely, the fraction is the cosine of Latitude 40° times the velocity of the Earth at the equator." She drew some angles and jotted down formulas. "So over twelve hundred kilometers per hour. Not enough for a circular orbit at this height, but it's enough so we *will* miss the Earth as we fall back. Our velocity perpendicular to gravitational pull at this height is is over one thousand kph, more than enough to put us into an elliptical orbit."

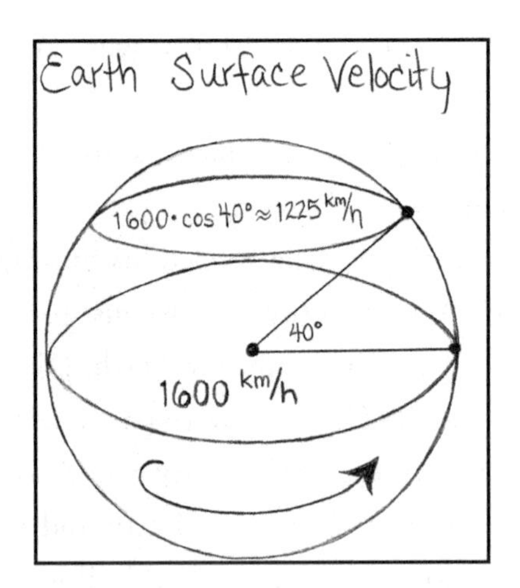

From the notebook of Catherine Cavor

"So if we will miss by falling back towards Earth, how do we get home?" Noah asked.

"This spacecraft has no other means of propulsion except gravity. We can either push or pull. So if the Earth pulling us will not work, we need to find some other mass to push against to return home."

"But what can we push against?" asked Dana. "This is outer space."

"I am afraid to say it, but at this point, the most secure method to return is going to be for us to go to the Moon first."

Dana didn't think it felt quite right that they needed to go all the way to the Moon to cancel out velocity they had picked up from the Earth's rotation, but she trusted Dr. Cavor to know what she was talking about.

"So we are making an unplanned trip to the Moon in a completely untested spaceship," said Dana. "Wow. I always said I wanted to be the first girl on the Moon. I guess when they say 'careful what you wish for' they're really not kidding!"

"And you, Mr. Bosco? Did you ever wish to be the first dog on the Moon?" said Dr. Cavor. Bosco looked over at the mention of his name but made no comment.

"What about that outer airlock door?" asked Noah, looking worried. "Are we really in more danger with it open?"

"I think we are ok," said Dana. "We are well past the atmosphere already, and I am not hearing any more creaking from the frame. It's holding so far."

"I have an idea for closing that," said Dr. Cavor. "We are going to rotate the spacecraft, and let gravity close the door for us."

"How do we rotate the spacecraft?" asked Noah. "We don't have any jets."

"For want of a better image, think of it as an immense hamster ball," said Dr. Cavor, "and you and Dana will be the hamsters. I want you two to start over here." She pointed to a cell of the dodecahedron next to the airlock. "As soon as I close all the blinds and we are in zero gravity, I want you two to make your way up that wall towards the top, as fast as you can, then down the other side. Stop when I say so."

"I get it!" said Dana. "As we climb one way, the spacecraft spins the other way."

"Right," said Dr. Cavor. "Stop when I tell you to."

Dana and Noah took their positions by the airlock and waited for Dr. Cavor's signal. Bosco, excited that something was happening, joined them, his tail thumping against a window.

Dr. Cavor positioned herself at the top of the spacecraft with the brains that worked the shutters. "Ready?" Dr. Cavor said. "Three... two... one... go!" Shutters closed and Dana felt her stomach drop a little, like on the big hill of a roller coaster. The kids scrambled up the side, grabbing struts to propel them along. Bosco did his best to follow them, without the obvious advantage of opposable thumbs.

"Now stop!" Dr. Cavor said, and the shutters at the bottom opened again. A fraction of their weight returned, and the children grabbed struts as their bodies naturally reoriented to the new up and down. Bosco slid down one side of the Beetle, legs scrambling to find purchase. He ended up in his previous position down by the airlock, still happily whacking things with his tail.

Dr. Cavor manipulated the controls and the whole biosphere seemed to spin and rock for a while. "That did it," she said.

"So, how did that work?" asked Dana. "If we are shielded from gravity in here, why would the hatch fall back into place?"

"The inside of the airlock is not shielded. But also, our sensations of weight are always going to be a little peculiar in here. For example, right now you feel some weight, even though the gravity shielding is preventing the Earth from pulling you. That feeling is not gravity. It is acceleration. The earth is pushing on our neg-grav core and accelerating the Beetle. You are being pushed by the floor of the Beetle, and that sensation mimics that of gravity."

"Weird," said Noah. "This will all take some getting used to."

"It certainly will," said Dana. "But in the meantime, Dr. Cavor, I think you have some explaining to do!"

CHAPTER 16 — THE LUNARTICS

Dr. Cavor laughed. "*I* have some explaining to do, do I? Are you sure you and Noah are not the ones who need to explain to *me* how it is that we are flying to the Moon powered by exotic matter that, as far as I know, could only have been produced in my completely off-limits private laboratory? We are only in this predicament because you broke into my lab and stole equipment."

Dana look abashed at Dr. Cavor's accusation and started to explain, sheepishly, "Well, we needed to test the strength of the transparent polycarbonate for the windows..."

But Dr. Cavor didn't actually look angry. Her strange, opalescent eyes sparkled with amusement behind her glasses. "It is certainly reasonable for you to be interested in how there is a machine in my lab that can alter matter in ways about which modern science does not have an inkling."

"Yes," Dana said, "That. Also, it would be nice to know how the machine works. What is the science behind all this? How are you so far ahead of everyone else to be able to invent something like this?"

"Let me start a little lesson. Settle in for story time, children. We might as well become comfortable in our surroundings, as we will be here for several days before we are able to reverse course."

"Days?! How many days?" Dana felt anxiety twist in her stomach. Considering they had just *launched themselves into space*, she knew she should have bigger concerns, but... "Are we going to be back in time for the contest judging?"

Noah stared at her, mouth open. Dr. Cavor blinked, slowly, twice. Bosco wagged his tail *wumpwumpwump.*

"I guess that's not really that important right now," she mumbled, twisting her hands together.

Dr. Cavor smiled and her eyes twinkled. "You certainly have every right to be proud of your project. You have both worked very hard, and I hope you will be able to reap the rewards of those efforts. I believe that yes, we should be able to return before the contest judging. It should take approximately three days to get to the moon, which puts our lunar landing on Saturday. It will take approximately an equal time to return to Earth. You should be able to land in place outside the gymnasium on Wednesday, barring any unforeseen events."

Noah just shook his head and mumbled something about priorities.

"Now children, let's start our lesson. Your first question is rightly: why do I have access to such advanced scientific theory and technology?"

Dana opened her mouth to say something but Dr. Cavor raised her hand. "We will conduct a separate lesson on how the machine works at a later time."

Dana closed her mouth.

"First, let me ask you this. Who do you think is better at math and science, you or Noah?"

"We like to do different things. I like astrophysics and electronics, and he prefers microbiology and genetics."

"Very good," said Dr. Cavor. "And who do you think is smarter, girls or boys?"

"Neither," said Dana. "I think that it's a case-by-case thing. You can't assume anyone is smarter based on their gender."

"Very good," said Dr. Cavor. "So if both those things are true, how do you account for the fact that almost every famous scientist, mathematician, and inventor that you have ever heard of in all of history is male?"

"Cultural factors," said Dana quickly, having thought about this topic quite a bit before, like every time she had sat in Principal Peters's office and he had told her naughty little girls didn't belong in a science lab. "Historically, most cultures have had ideas about roles for men and women, and those societal roles meant women's participation in such work was banned, discouraged, suppressed, or ignored."

Noah said, "There's also the issue of who gets credit. Female scientists have done a lot of work that wasn't recognized. Or they didn't get the same level of recognition that their male colleagues received."

He continued, now fully interested in the lesson. "I know that James Watson and Francis Crick are the names people associate with the discovery of DNA. But all they did was create the first models of its double helix structure. What most people don't know is they did this *after* looking at the first X-ray pictures of DNA taken by Rosalind Franklin. Her pictures were shown to Watson and Crick, without her permission, by a colleague who worked in her lab named Maurice Wilkins. Watson and Crick only 'discovered' the double helix structure after looking at her pictures of it. So, Watson, Crick, and Wilkins all shared the Nobel prize, but Franklin died of ovarian cancer from too much X-ray exposure. And that's just a famous example we know about. I figure a lot of men have gotten — or taken — credit for work that was done by women, and we'll never even know."

"Both excellent points," Dr. Cavor said. "But I happen to know there is another additional reason that great female minds have not been as often recognized."

Dr. Cavor paused dramatically and Bosco made a weird noise and kicked himself repeatedly in the head with one of his back feet, apparently trying to scratch at an itch.

"Let me begin by telling you a story. It is not 'once upon a time,' but it is almost as fantastical as those fairytales. This is something that you would not have heard in your schools.

"In the later part of the eighteenth century, in and around Birmingham, England, and in some places further away through correspondence, a loose society of geniuses was formed. They called themselves the Lunar Society, because they usually met on nights of the full moon. The members were not all considered scientists by the Royal Society of London, who looked down on some of what they studied as being the realm of mere tradesmen and tinkers. But this group, drawn together as fellow seekers of knowledge, did not care.

"They mocked their own outsider status, calling themselves the Lunartics. This joke name, like so many other things they wrote, said, and did, also had a secret double meaning. The 'tics' part of Lunartics stood for their four Disciplines: Technica, Industria, Calculatia, and Scientia. Today we know these as Technology, Engineering, Mathematics, and Science. The STEM disciplines."

Dana looked at Noah as he stifled a yawn. He was never as interested in history as she was, except maybe when it was the history of bioscience discoveries, but he was doing his best to appear politely interested.

Dr. Cavor, either unaware of or uninterested in Noah's failing attention, continued. "But the most shocking and outrageous thing about their society is that they held as a truth that women had as much right and ability as men to join in this quest for knowledge. In fact, they freely accepted women and men of any race or culture into their circles, judging their fitness for participation only by the quality of the thoughts in their minds. While this idea is more accepted today, at the time, such free thinkers were not welcome in society. In the end, they were too public with their unpopular ideas and, as many people whose ideas and attitudes are out of sync with their time have been throughout history, they were persecuted. Mobs with torches and pitchforks arrived, burning members' houses, and driving them out of town, and underground. What members were left were forced to meet clandestinely." Noah perked up again at this mention of secret societies.

"When this happened, some of the members kept up their studies but published more socially approved work in more appropriate circles. Those who did not feel welcome in those circles, most particularly a number of very intelligent women, continued on in secret. They started doing their own research, having their own meetings, and sharing their great discoveries only amongst themselves. As the Lunar Society grew, it mostly did so hidden in plain sight, under the disguise of girls' boarding schools, with a stated mission of producing well-educated and proper young ladies. Such schools were established in many countries around the world. Schools like mine. Where the best and brightest of students can be recruited, educated, and encouraged."

"And there are no male members?" ask Noah.

"There have certainly been a few notable male members, both at the beginning and then later. Mostly husbands and sons brought into the fold, but our secret history is as full of great breakthroughs by women as conventional history books have been dominated by men. Over the years, many a well-known male scientist has had a wife, sister, or daughter in his household who was surreptitiously doing research far beyond his own level, perhaps even giving him the occasional hint and nudge in the correct direction."

"Well, Professor Moreau is a man and he has certainly been doing some strange advanced research," said Dana. "We've seen his lizards."

"Ah, yes," said Dr. Cavor. "His eyes and ears. He likes to keep track of important things."

"Are the kids in the STEM contest 'important things'?" asked Noah. "All the kids in the contest have seen those lizards around their houses, as long as they can remember. All of them but me, for some reason."

"Young geniuses are important to all of us at the Cavor School. We try to find them and teach them. Also, protect them. Often from themselves. Genius is not always survival-oriented. A charging tiger will send a normal person running and climbing a tree, while a genius might stand there distracted, thinking about the evolutionary reasons for orange stripes. The lizards may be a little strange, but try to think of it as a concerned adult keeping an eye on all of you."

"But not me?" said Noah.

"He may have missed you somehow. Or perhaps he utilized another method that a biology enthusiast such as yourself would not have noticed as being unusual. In any case, he is skirting the rules with those lizards. The Lunar Society has a few important rules, one of which is 'Only one body allowed,' or at least only one at a time. He has not been censured so far because they are not thinking creatures on their own — they are not actual copies of him. But he circumvents the active measures we have set up at Myrtlegrove

Manor to prevent such things near the school. I have had words with him about how that might be dangerous, although I have not reported him to the council.

"But we do try to keep things safe. From the very beginning, the Lunar Society research has been far ahead of its contemporary counterparts, and we consider it our responsibility to make sure new technology is used in beneficial and responsible ways. We try to prevent dangerous technology from being unleashed on an unprepared world."

Noah said, "Wait, I'm confused. If the women have always been ahead and are watching over the science that men do, making sure it's responsible, doesn't this mean women are smarter than men?"

"No," Dr. Cavor laughed. "Well... at least, not necessarily. We are not ahead because women are smarter. We are ahead because we took a shortcut."

"What?" Dana interjected. "How can you take a shortcut? Science always builds on previous science, a little at a time. Doesn't it?"

"Yes," Dr. Cavor said. "Perhaps 'shortcut' is not the right word. Let us say that if Lunar women have seen farther, it is by standing on the shoulders of giants,[72] of whom the men were unaware."

Dana was about to ask more about these unknown giants of science, but Dr. Cavor suddenly asked, "Might I ask, how does one use the, ahem, facilities? I am feeling in need of a break."

The children allowed their momentary embarrassment to distract them and they let Dr. Cavor get away with changing the subject.

"Oh," said Noah, "That's not, um... exactly easy... we used suction systems for collecting wastes for re-use in the cyan-aid tanks. We were saying that this method would actually be useful in zero-G in a real space flight. I guess now we'll get to test that out. But it's... well... I guess Dana can show you."

---

[72] "If I have seen further, it is by standing on the shoulders of giants." — Sir Isaac Newton (1642–1726)

"Well," said Dr. Cavor, "If it will not be easy for me, I expect that it will be even more difficult for Mr. Bosco."

Bosco, hearing himself being talked about, looked up expectantly and started banging his very hard tail against one of the support struts, causing the whole habitat to ring like a somewhat muffled bell. Dana and Noah, however, were much less excited, and both groaned at the thought of figuring out a regular procedure to deal with the bodily functions of a large dog (and a regular-sized fox) aboard a very small spacecraft.

The next order of crisis management to address was the issue of their air supply, starting with Dana asking Noah a very important question. "If you designed the air recycling systems to be used by two children, how do they hold up to that plus an extra adult, a large dog, and a fox?"

Noah looked startled. Then he said, "Well, the good news is I was calculating for two average-sized adults. But I was on the edge for that, based on the stats I found for average daily oxygen values. I think we might be ok, but only if we try to keep calm and mostly immobile."

"Sure," Dana said, dripping with sarcasm. "First you tell us that we might run out of air and die of asphyxiation while hurtling to the Moon in an untested spacecraft, but then you tell us to 'keep calm.' What an excellent idea! That will solve all our problems! Don't worry, be happy! I can't believe I didn't think of that! It's amazing how calm I'm already feeling — "

Dr. Cavor interrupted the beginnings of what was sure to be a magnificent Dana-esque diatribe. "I believe I can assist with that. We can go a level below 'calm and sedentary' with the application of some Tibetan yoga and self-hypnosis techniques. I can show you how the Devas Monks taught it to me. I can do a very deep trance, but just say my name and I will come out instantly. Now, make yourselves comfortable."

They each positioned themselves as comfortably as possible inside the structure of the spacecraft. Dr. Cavor instructed them to breathe deeply and clear their mind of all thoughts. Dr. Cavor was a real master of these meditative arts. She went into such a deep trance state that she didn't seem to breathe at all. Bosco climbed into her lap, which didn't seem to disturb her in any way. But Dana and Noah followed shortly behind her, like they were both naturals. It was as if they had done this before or it was coded into their genes, and as she slipped into oblivion, Dana realized it was a lot like the trick she did to put troubled thoughts out of her head to calm herself. This was just a matter of putting all thoughts aside. Soon they were in a deep meditative state in which, Dr. Cavor had assured them, they would be consuming very little oxygen.

Sometime later in the trip, not knowing exactly how long it had been, Dana came out of a trance, or maybe actual sleep. She was cuddled up against Dr. Cavor, who was immobile as a statue, and Noah was on the other side, also motionless. Bosco was still curled up in the teacher's lap. Dana's mind began to race, and she felt anxiety rising in her throat. It was hard not to think about the cold emptiness of space on the other side the relatively flimsy plastic-windowed, aluminum-framed, rubber-gasketed structure they were in. There was not a lot between their life-giving air and the vast expanse of nothingness.

Dana looked out one of the open-shuttered windows. It was full of gorgeous points of light, with more individually discernable stars than one could make out on the darkest night in the deep deserts on Earth. Seeing the stars, without the lens of the Earth's atmosphere, was a truly spectacular experience.

She was elated, but also terrified. It was a real possibility that it could all end here. That she might die. She wouldn't win the scholarship. She wouldn't learn the secrets of Dr. Cavor's Lunar Society. She wouldn't become a famous scientist. She would float away into the void. Her parents wouldn't even know what happened.

That last thought was certainly a terrible one, but she felt herself somewhat comforted by another one: *I'm here. I'm in space. I've done something extraordinary. I'm having the kind of adventure that very few will ever get to realize.*

Sure, I'm scared. This is a scary thing to be doing. But I wouldn't trade it if I could. If I could be safe on the ground, at home in my own bed right now, by wishing on these stars, I might make that wish. But if I were home, if I were looking at a much less clear and bright group of stars from the ground right now, I might be making a wish on those same stars, asking to be here where I am right now.

She looked away from the stars and considered how strange the lighting inside the Beetle was. Pitch black with bright stars, but she knew the brightest sunlight possible was on the other side of those slats there, and the biggest, glowingest, fullest Moon below them. She also saw some of the cyan-aid's UV grow-light coming from between the slats of the Beetle-heart. She pried back one of the slats to see how it highlighted tiny flecks of dust in the air, making them glow.

Then, out of the corner of her eye, she saw another faint glow. It was on the back of Dr. Cavor's hand, a few glimmering lines protruding from under her sleeve.

Dana reached out and pulled back the fabric to reveal an intricate shape on Dr. Cavor's hand and wrist. *It must be invisible in normal light,* she thought, but the UV LEDs[73] were making the ink glow bright yellow. It looked a bit like a cartoon stick-figure, with a strange hairdo or maybe large crescent-shaped ears, standing on a small rock — or moon or asteroid — her arms, legs, and smile all ending in arrow-like points.

This was weird, but not by far the weirdest thing to happen that day. She'd ask Dr. Cavor about the invisible tattoo later, when everyone was awake. But right now she was yawning, and it was time for resting and conserving oxygen. She settled back against Dr. Cavor and stroked Bosco's head and tried to fade back into a trance or maybe sleep. When it didn't

---

[73] *Light Emitting Diodes* are electronic components that convert electricity to a specific color of light.

work right away, Dana reached into one of her pockets and pulled out an expansion pack of Elementalist cards she had been carrying. She went through the cards, one at a time, looking at the art and attributes of each elemental. It was a good way to memorize basic chemistry knowledge and also try to take her mind off all the things that were making it hard for her to relax and sleep. Counting chemicals, Dana drifted. So many strange things had been happening to her. Her eyes got heavier, and thoughts of these events seemed to coalesce and split, recombining in new dimensions and strange ways. She looked at the card of a tin elemental. It was not Stanus, the one she had beaten Noah with seemingly a million years ago now. This was Stana, a lower atomic weight isotope of tin with a picture that looked like a female version of the Tin Woodsman. "Bosco, I've got a feeling we're not in Kansas anymore," she mumbled, as much to herself as her faithful pet.

She put the cards away and settled into a deep sleep...

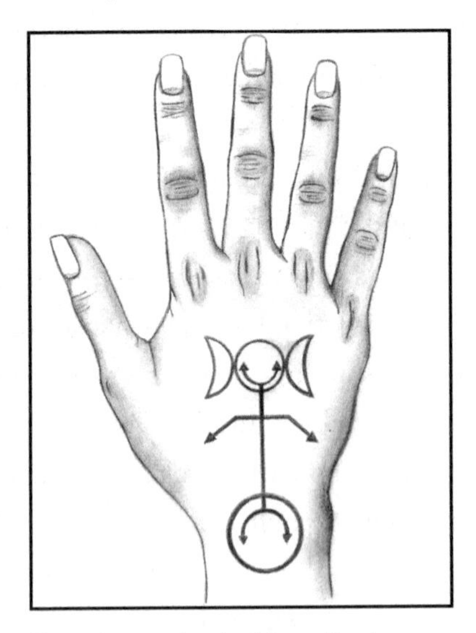

From the notebook of Dana Dash

CHAPTER 17 — HOLE IN ONE

In Dana's dream she was walking, Bosco ahead of her, on a golden path. It stretched up a hill before her, so she could see its pattern of crossroads, curves, circles, and strangely pointed ends. Noah and Dr. Cavor were there too, each holding one of her arms. At first it didn't seem strange that Noah was covered in orange fur and that Dr. Cavor was made of gleaming metal. Then she took another step and looked at her feet and saw her silver shoes against the yellow bricks of the road.

"Oh!" she said. "This is a dream!"

She looked at Noah again and he was, indeed, a timid lion. He had a tail and mane and several long stiff whiskers protruding from his upper lip to each side of his nose. And Dr. Cavor was a tin man — er, person. It was her ordinary face, but metal all over, with her usual glasses balanced on a silvery nose. Instead of her black umbrella, she carried an axe, but with the same black double helix handle. With herself as Dorthy and Bosco as Toto, the whole whole Oz gang was here.

Almost. Wasn't someone missing? She looked back and forth, and then finally, turned to look behind her.

The Scarecrow was there.

Following behind.

It was like the vision in Dr. Cavor's office, like the thing riding the deer, with the same hat, but with a full body and detailed face made from rags and leaves, burlap sack cloth and trash. Things wriggled underneath the trash, animating its arms and legs and its head too. A horrible moving caricature of a human face sculpted in garbage.

It reached out to touch her...

From the notebook of Dana Dash

CRACK!

There was a sound like a gunshot that she seemed to hear and feel through the sides of the structure where she touched the walls and deck, and even through the body of the woman she leaned against. The sharp sound was followed immediately by the whistling sound of precious air escaping, and then the sound of liquid spilling in a constant stream. Bosco bounded off Dr. Cavor's lap and began squeal-barking madly.

"Ohhh nonononoooo!" she yelled, "We're hit! Something hit us! Noah!"

"Hrmm... What?" Noah replied groggily, having been roused from deep slumber. "What's going on?"

"A small meteor I think. We're losing air. Quick! Help me find the hole. And I need something to patch it with."

Dana began casting about with eyes and ears at full alert, trying to determine where the whistling noise was coming from. Her ears were already starting to feel like they needed to pop to equalize pressure.

She saw Bosco was on the other side of the ship, his paws up on the wall, barking at something like he really meant it. This was not his usual excited squeal-bark. She made her way over quickly, and found the whistling hole. She covered the hole in the polycarbonate window with her hand without thought. The whistling stopped, and Bosco stopped barking. She looked at some tiny fractures radiating out from where her hand covered the hole. At least the whole spaceship hadn't popped like a balloon after a pin prick.

Her hand felt uncomfortable and she pried it up to find an already painful red welt forming. Bosco resumed barking again.

"Owww!" She said. "I found the hole, but it hurts to keep it plugged." She replaced her hand in a slightly different position to spread out the damage it was doing to her palm. Bosco quieted down, still alert.

Noah was more awake now. "My tanks are draining out!" he yelled. He went scrambling up to his tool area. "Hang on. I brought some plumber's putty."

"Come do the air leak first!" she insisted, shifting her hand again in response to the pain. Thinking quickly, he broke the stick of plumber's putty in half and tossed her one of the pieces. She caught it with her free hand.

He moved to his cyan-aid tanks, quickly locating two streaming holes and filling them. While he was working on the tank, Dana removed her hand and applied some putty too fast for Bosco to object to the renewed whistle. It held for a moment, but then it started to dimple in the center as the putty was slowly sucked out through the small hole.

"Well, that's not going to hold," she said to herself and Bosco.

She went up to her supply area and found some of the spare gasket material. She had long strips cut to match the lengths of the Beetle's edges. She also had the circles of rubber like she had used at each of the vertices of the structure. She grabbed several of these rubber circles, and the glue she had originally used to attach these gaskets at each corner.

The plumber's putty failed with a sucking *schhhwupp*, and the whistling soon resumed. Bosco had barely gotten a chance to start barking at the leak again before she got the first layer of glue and rubber down to silence both him and the puncture. By the time she had the third layer of patch applied, Noah was next to her.

"Looks good," he said. "Will it hold?"

"Should be fine," she replied. "The hole isn't large. It's like a bullet hole."

"Like a bullet..." said Noah. Dana followed his gaze as he looked from the hole in the outer hull of the craft, which had punched neatly through the outer shutter and the window, to the matching hole through the inner shutters and the polycarbonate faces of the water tank.

Dana felt a sense of growing dread as the children moved around opposite sides of the craft to find the next hole. The mental connect-the-dots created a straight line through all the holes and continued on right to where Dr. Cavor still sat. Her eyes remained closed and she was unmoving.

"Oh no!" said Noah, as they moved to either side of her.

Dana reached out and pulled at her arm. "Are you ok?"

Noah took her other shoulder in his hand and shook her lightly. "Are you hurt? Please say something!" She fell right onto her face, unmoving.

Can't scream. Can't spare the oxygen, Dana thought. She only said quietly, "She's dead. It lines right up with center of her chest. It didn't go through the far wall behind her."

Bosco whined and sniffed at the face-down woman.

"She stopped it," Noah said. "If there had been two holes in the spacecraft, you might not have been able to patch them both in time. She may have saved our lives by taking the equivalent of a bullet in the chest."

Dana pushed Dr. Cavor's lifeless body upright and Bosco tried to climb back onto the woman's lap. Dana stopped him.

"No Bosco!" she scolded and grabbed his collar. "You'll get blood on you —" She stopped. Then, "Where is the blood?"

"Dr. Cavor?" said Noah.

Dr. Cavor opened her strange eyes and sat up abruptly, adjusting her glasses on her nose. "Oh, hello children. And Bosco." She welcomed the dog back onto her lap as Dana released him. Bosco sniffed at Dr. Cavor's chest as she looked back and forth between the expressions on the faces of the two children.

"Oh, I am terribly sorry," she said, "I must have frightened you. I warned you that you might have to say my name to wake me up before I was ready, yes?" There was a strange buzzing sound to Dr. Cavor's voice as if she had a bad chest cold.

"We thought you were dead!" said Noah.

"There was a meteor," said Dana. "It punched a hole in the Beetle. And we thought it got you too." Dana was examining Dr. Cavor's blouse. Was that a hole she saw? Had it been there before?

"Oh my!" Dr. Cavor buzzed. "Well, technically a meteorite then. You know they change name from meteor to meteorite when they survive entry into the atmosphere, and we, fortunately, still have atmosphere here. You both must have acted quickly and bravely."

Dana was simultaneously annoyed at the nomenclature correction and elated to be praised by Dr. Cavor. The woman could not have been more distracting to her train of thought if she was doing it on purpose. *Perhaps she is doing it on purpose.*

"Does the atmosphere inside a spaceship really count?" she asked.

"Oh, I don't know. Call it whatever you prefer." Dr. Cavor said, coughing, clearing her throat, and rearranging her shirt collar and sweater. Something went ***CLANK*** below her and she reached down into her lap, under Bosco, and pulled up something between her thumb and forefinger. It was a flattened lump of metallic indigo rock. "This looks like our stowaway here. Thought you would hitch a free ride with us, eh?" she addressed the rock, then handed it to Dana.

Dana examined the metallic meteorite and wondered how many thousands, millions, or billions of years that this rock had been spinning around the solar system to finally happen to run into them now. And she thought about how close it had come to killing them all. Dr. Cavor didn't seem to be worried about anything, and that attitude was somewhat contagious. It felt like nothing could go wrong with her there. But it was also hard to look at a small rock, that had almost ended their lives, and believe that any of this was even remotely safe.

"Lucky shot," said Noah. He was looking at his cyan-aid tanks. "None of the grow light tubes were damaged. It's almost a miracle that it could punch through one side and out the other without hitting any of the internal structure. If it had shattered any of those tubes, I don't know how we could have made repairs easily out here."

Dana slipped the rock into her pocket and said, "I'll glue a couple more rubber patches over the holes you plugged."

Dr. Cavor said, "It was almost time for me to wake up anyway. We are over halfway there. I need to apply a course correction soon. We need to stop pushing off the Earth and start pushing off the Moon so we will not come in too fast and end our journey by adding one more crater to the moonscape."

"Wow!" said Noah. "How long were we asleep for?"

Dana checked the time. "Wow indeed. We've been out of it for well over a day. That meditation thing really works! How do we steer this spacecraft next? A regular rocket might have another stage burn..."

"One moment, Dana. Let me find a pen and I will show you and Noah the calculations for piloting this craft." Dr. Cavor patted Bosco affectionately on the head and got up. She went to Dana's storage area and rummaged around for a few minutes, then was quiet for a few minutes more.

"Dr. Cavor?" Dana called out. "Did you find what you were looking for?"

"Yes," Dr. Cavor said. The strange buzz was now gone from her voice, and she was speaking normally. She came back to sit down and sketch some diagrams and equations in her notebook as the children watched. Leaving Dr. Cavor to her calculations, Dana went to get a few more patches for the punctured cyan-aid tanks. She went to the drawer where she had left the rubber gasket material. She thought she had seen three more of the round pieces there, when she had been scrambling to patch the outer windows. But she had been very distracted then. Now she could only find two of them. But that was all she needed. She grabbed the glue and began making the necessary repairs to the oxygen production tanks.

"Well," Dr. Cavor said. "I have the rest of our course figured out. I am going to close all the shutters and coast us for a while. We will not feel any gravity for a bit. We we will need to rotate the spacecraft so our landing plate keel is pointed towards the Moon before we can start braking against the Moon to match its orbital speed around the Earth."

There was the previous dropping feeling in her stomach when all the shutters were closed, and in one way, it was only a very minor change. As they had gotten farther and farther from the Earth, the push they were getting from their home planet had decreased considerably, so they hadn't really lost much weight. In another way, however, it was a totally new thing. Even a little gravity allowed a normal feeling of what was up and what was down. With no gravity to guide the mind, one's mental frame of reference could shift without warning. Like an optical illusion that changes from being one thing to another and back, without warning Dana would instantly go from feeling she was lying on her back to thinking she was falling face forward, and her hands would instinctively jerk up to catch herself.

But after a reasonably short period of time, she was enjoying it. She could spin in the air like a top or tumble in an endless somersault. She and Noah took turns trying to outdo each other with clever moves. It was a great game.

That is, until Bosco threw up.

They had finished chasing down the last of the dog barf, sucking it out of the air with the tool they had improvised for bathroom use, and wiping it off of windows and struts with rags, when Dr. Cavor directed them to action.

"Time to rotate the Beetle. The same way we did it when we closed the hatch. Start there and end there." She pointed out the course they would take.

When they were done, Dr. Cavor, still at the controls, said, "Something is wrong."

"What's wrong?" both of the children asked in unison.

"The Earth-side outer shutter was supposed to re-open — but to have Earth pull on us, rather than push — to help slow us down so we can match lunar orbit."

"It didn't?" said Dana, "Let me see the brain."

Dr. Cavor said, "The control program says it is open, but it is not. Children, move around the cabin to my other side."

The two children complied.

"That was enough to give me a different face earthward, and... no, that is not opening either. Duck under me and go around again..."

"Are we in trouble?" Noah asked as the kids finished their second trip around the craft.

"No," Dr. Cavor said. "No, it is all right. That one opened. I believe things are in order again. The ones that failed are now towards the sun, and we want to keep the shutters sunward closed regardless, so we do not overheat.

"But we would have been in some trouble if we had not been able to get our brakes working right for much longer. This is not an easy spacecraft to pilot. Rocket pilots have jets that they can point in any direction, at any time, to correct their course. We can only pull towards or push against far-away massive objects to alter our course. And when there is side velocity relative to that push or pull vector, and there almost always is, we will not actually move away, or towards, that object. Instead we will move in an orbital curve that takes some serious math to calculate.

"So, to go from the Earth to the Moon, we need to push off from the Earth until we have enough velocity to travel the distance in a reasonable time, but then at some point, we need to push on the Moon as well, both to reduce our velocity away from the Earth and to match velocity with the Moon. That way, when we reach it, we will have attained the same orbital side velocity it has relative to the Earth. We need to have matched its orbit at exactly the same time as we arrive at its orbit, and as it has reached that same point to meet us. Timing and angles have to be nearly perfect. Imagine playing golf and trying to hit a hole in one with each shot. One wrong calculation, and we will remain in orbit longer than we can remain alive. Or, we could approach too rapidly and — well, children, I don't think I need to explain the consequences of one object colliding with another object. Suffice it to say, splattered on the Beetle's windshield is not my preferred way to exit this Earthly life."

CHAPTER 18 — EXTRA-VEHICULAR ACTIVITY

Before they settled in for the next leg of their journey, they all took a moment to check out the systems and deal with their biological needs. Dana and Noah had anticipated dealing with bathroom procedures from within the confines of Earth's gravity; working out the difficulties of in space was, to say the least, tricky. Fortunately, the shutters on the walls of the airlock provided some privacy for each of them to work it out as best they could. Food was the next issue; Dr. Cavor said she was fine without food, but encouraged Dana and Noah to eat to keep up their strength and mental activity, even if their appetites had stayed behind on Earth. They munched on their protein bars dutifully, but Bosco seemed genuinely happy with his ration, since human food was not a normal part of his diet, except on the occasion he found a morsel dropped from the table.[74]

Noah ran some tests on the oxygen farm tanks and fed the acquired biological waste materials into the parts of the system designed to contain and recycle them. Chip had been pawing at his water dish again, and Dana, knowing they needed to conserve their drinking water, dropped the meteorite into his dish with the other stones. *Hopefully*, she thought, *that will weigh it down a little more. And at least I'll know where it is when we land.* It was a real meteorite, and Dana had every intention of keeping it as a souvenir from space. She didn't want it being tossed out as an ordinary rock when they landed, and with all the space-somersaulting they were doing, it was certain to fall out of her pocket.

Dana was still concerned with the stuck outer shutters. With Dr. Cavor's help, she took another look at the problem. She found that the

[74] The three-second rule, as applied to dogs, is closer to a three-week rule.

rubber patch she had applied to the small meteor hole was quite close to the vertex where three of the pentagon-shaped windows met, two of which had been the ones that Dr. Cavor had been unable to open.

"Try opening number twelve," Dana said to Dr. Cavor.

"For only a moment," Dr. Cavor said. "Turn your head. You do not want sunlight directly in your eyes, especially with no atmosphere to reduce the ultraviolet light." Then after a short pause while Dana turned her eyes away, "No, that one is not opening either."

Dana turned back to look at where the wires went through the strut. "Ok, so those three are on the same bundle of control wires," Dana said, "The meteor hit could've damaged them all at the same time. But I don't get it — the outer shutter should be springing open if the power wires are broken. But it's stuck closed."

Dana gave the window near the meteor hole patch a medium hard whack and all three stuck shutters popped open at once. The full light of the sun, with no atmosphere to soften it, immediately blinded her.

"Owww!" she called out, closing and covering her eyes.

"Hey!" yelled Noah, "Close those shutters! Quickly! I don't know how much of that level of light the cyan-aid can handle."

"They still are not responding," said Dr. Cavor. "But now they are stuck open instead of closed."

"Close the inner shutters," said Dana. "The ones on the tank frame that are in the sunlight now."

"I have it," said Dr. Cavor, and all the inner shutters on that side of the Beetle-heart closed.

Everyone moved down and around the air recycling tanks to be on the opposite side from the blinding light now flooding one side of the little spacecraft. Dana's vision had a pulsing black and red spot in the middle where the sunlight had attacked her retina for only an instant.

"Well, that's not good," said Noah. "How do we get those closed? The bacteria in the tanks needs to be kept in a certain temperature range to stay active and alive. The tanks have heaters but no cooling system, and I can already feel it getting warmer in here."

Dr. Cavor said, "Human beings and animals also need to be kept in a certain temperature range to stay active and alive. We need to fix this problem very quickly."

"Can it really get that hot?" Dana asked. "I thought space was cold."

"That is a common misconception," Dr. Cavor said. "Space is a vacuum. And while it might technically have a very low temperature, because there is nothing there to hold any heat, that does not mean it is very good for cooling us off. In fact, the opposite. Vacuum is one of the best insulators there is. That is how you can still have warm soup at lunchtime when you cooked it and put it in a thermos early in the morning. The walls of the thermos are hollow and mostly evacuated. That near vacuum keeps the heat energy in the soup from escaping.

"Right now, the sun is heating up the air inside of our thermos bottle, and the vacuum of space is preventing it from getting out. Your outer shutters were perfect for preventing this. They are mounted so they stand off the surface of each window enough to open and close. There was always a layer of vacuum between the hot sun side of the closed shutters and the windows. But with the shutters open to the sun, we are all going to be in very, ah, hot soup, and soon."

"Ok," said Noah. "So, what do we do? Can we rotate the open shutters away from the sun again?"

Dr. Cavor said, "Not without disrupting our flight plan. If we do not have the landing plate pointed at the Moon as a keel, and pull from the open shutters pointed towards the Earth, we cannot reliably keep the craft oriented properly to both slow us down and keep us heading towards the Moon. And we very much need to be slowing down right now, relative to the Moon. I was not making a joke about the possibility of making a new crater if we came in too fast."

Dana interjected, "The wires that go to those servos are broken outside. The only way to fix them will be from the outside. We have a spacesuit we made for this sort of problem. Although it was meant to be used to make external repairs on Earth — I never thought I would actually be using it in outer space."

"Now wait a minute," Noah said. "It's not safe for you to go out there."

"Safe? Is *any* of this safe?"

"I don't want to go either," said Noah, "and I know you're better with wiring stuff. I was hoping there could be another way."

Dr. Cavor said, "Dana is right. It has to be done, and it has to be done quickly. But we are under deceleration now. It is not significant, but as you can feel, we effectively have some gravity. Down is down again. You will need to climb up the outside of this spacecraft, and if you fall, we could lose you. I do not suppose you brought any sort of rope to tie to yourself, did you?"

"No," said Dana. "I have some extra wire, but not a lot. We didn't plan on any actual space walks. When we pictured going outside to fix a shutter, we assumed we'd have a place to stand with, you know, gravity. Maybe a ladder borrowed from the custodian." She looked at Noah. "Let's do this. Get the space suit."

Noah opened the drawers where they had stuffed the space suit and passed it to Dana. As she struggled into the suit and zipped it up, he detached one of their two compressed air tanks from a bracket and secured it to the harness on the back of the space suit. Then he attached the hose to the helmet and handed it to Dana.

"Are you certain this is airtight?" asked Dr. Cavor.

"It's a diving suit. It's actually really good," said Noah. "We tested it at two atmospheres on the ground. Dana built the scuba gear regulator into the helmet. Each breath she takes will actually be exhausted from the suit as new air is brought in from the tank on her back. So it's not a closed system."

Dana added, "The whole spaceship isn't an entirely closed system either. We'll lose air every time we open the airlock. We have an extra air tank on board, but we can really only do this a few times. We didn't have a chance to install air pumps in the airlock. With no real vacuum involved, it didn't seem critical."

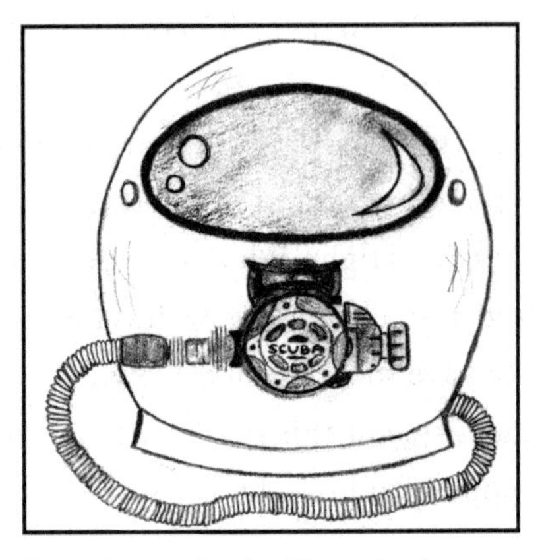

From the notebook of Dana Dash

"Let's not worry about that now," said Noah. "How will we talk to you for testing the repairs?"

Dana found the two ham radio handsets she had borrowed from Jayla. She said, "I put a bracket on the side of the pressure suit helmet to hold one of these. It wasn't a solution intended for use outside Earth's atmosphere, but it's held tightly to the side of the helmet. Hopefully there will be enough sound passed from direct contact. If that doesn't work, maybe I can press my faceplate against the window and yell?"

"No," said Dr. Cavor. "That will not work with the shutters. We can probably figure out a system with hand signals, if the radio will not work in a vacuum. Now, out you go. It is becoming uncomfortably warm in here."

Dana turned on both radios, put her helmet on and sealed it, then clipped the radio to the side of her head.

"Hello. Hello. Testing..." Dana's voice was barely audible from inside the helmet, but it came out loud and clear over Noah's radio handset. This produced an odd effect, because there was a slight delay between the two Dana voices.

"I'll need some tools," said Dana. "Noah, please get me the red wire stripper, the multi-head screwdriver, and some of those small blue wire nuts. Oh, and I'll likely need a crowbar."

"A crowbar? What for?" Noah asked as he went to get the tools.

"To open the airlock door against a couple thousand kilograms of pressure. It's only held closed against its seal by the internal pressure, remember?"

"Ahh, right."

Dana took the tools from Noah, securing some of them in velcro pouches on her waist. She entered the airlock and closed the inner door behind her. Then she inserted the crowbar under the seal to the outer door at her feet and began to pry. All she managed to do was lift herself off the ground.

"Not enough gravity in here," she complained. Then she walked up the wall so she was upside down and pushed off the central tank. She grunted loudly as she pushed the crowbar, and she heard a hissing noise, even through the suit helmet. After a short while the bar felt a bit looser and her space suit began to inflate to stiffness.

But then something was wrong. The suit was going limp again. The air was being sucked out of her lungs and at the same time she wasn't able to breathe in from the regulator. She released the crowbar and put her hand to the valve that was hissing away her air.

"What's wrong?" asked Noah.

Dana thought for a moment, and then said, "This scuba gear wasn't designed for outer space. All the valves are designed with the idea that the air pressure in my lungs will only be a little bit different than the pressure around me. I need my pressure gasket kit."

Dana pried at the inner airlock door which was now closed tight from the small pressure difference she had created by opening the outer airlock slightly. When it was open she had Noah bring her the box full of rubber pressure seal material. "I need to do two things. I need to wrap the tank valve to keep the pressure around it similar to that inside my space suit."

As she talked she was wrapping up the top of the airtank, where the hose connected, in layer after layer of the rubber strips she had used to seal the edges of the Beetle, tying them off tight. "And I need to do the same thing to the valve that lets air out when I exhale. Not close it off completely, but make it much harder to force air out, so it will work normally in the vacuum, not just spill all my air." After she finished this second task she said, "Ok. Here we go. Trying again."

She closed the inner airlock door and began to crack the outer one again. She saw the rubber strips around the tank regulator bulge slightly, but they held, and she was still able to inhale normally by finding the sweet spot on the breathing effort adjustment knob. When the escape valve on the mouthpiece built into her helmet began to leak again, she wrapped more layers of gasket tape around it until it only released air with the additional pressure from each of her exhalations.

She gave Noah a thumbs up as she opened the outer airlock door all the way. She was now in complete vacuum. Her makeshift spacesuit was surrounded by literally nothing.

"This is already seriously scary. Please be careful out there." Noah sounded worried.

As she made her way down through the outer airlock door, she realized that they had not designed the body of the space suit well for vacuum either. She was inside an inflated balloon that wanted to hold its shape, legs slightly apart and arms pointing straight out to her sides like a cheerleader spelling out a word with a "T" in it. It was a real effort to bring her arms down to fit through the airlock.

"I'm already planning some major design changes to this space suit," she said. "I feel as agile as a Thanksgiving Day parade balloon. We tested the suit for higher internal pressure, but that was before I added the air supply, so I was never inside to feel what it was like to move with it blown up like this."

"Are you going to be able to make it?" Noah asked.

His voice wasn't as audible as it had been with air helping to conduct sound to her helmet, but she could hear him, and he could obviously also hear her.

"I'm going to have to, aren't I." She spoke bravely but was feeling anything but brave. She looked down at the base plate she was standing on and saw something strange. Droplets of water had condensed on the plate from the air she had released from the airlock, but now they were boiling away again into the vacuum. Keeping her feet on the base plate, she held one of the support struts and leaned out to look down at the Moon. That was a big mistake.

Even the slight feeling of gravity she was experiencing, from the deceleration caused by the neg-grav water in the central tanks being exposed to the Moon, was enough to give her a sense of up and down. She was definitely looking down, and the Moon had never been down before — a long, *long* way down. The Moon was larger than she had ever seen in the night sky, even when rising or setting on the horizon, made larger by an optical illusion created by the atmosphere bending moonlight like a prism. But she barely noticed how spectacular the view was, because a part of her brain screamed at her that hanging a couple hundred thousand kilometers over the nearest ground was not a good place to be. She closed her eyes and grabbed the strut with both hands, vertigo washing over her, not even noticing for the moment how hard it was to make her ballooning suit bend to bring her arms together.

She must have made some sort of noise because she realized that both Noah and Dr. Cavor had asked if she was all right. She said, "Give me a minute — I'm trying to get a handle on this."

She tried to calm herself, using the yoga breathing that Dr. Cavor had taught her, until her heart was no longer pounding. She really wanted to climb back inside and forget this whole thing, but she knew that was not an option. She needed to fix this shutter problem soon, before it got any hotter in their little space capsule.

Dr. Cavor said, "I am going to open all the shutters I can, on the side opposite from the sun, so you can use the open shutter slats as hand holds."

Dana opened her eyes and found her way around to the landing strut on the shady side. Looking up she saw Noah and Bosco looking down at her, through the open shutters, from two of the outward sloping faces of the Beetle above her. Noah gave a little wave and Bosco was madly wagging his tail, his tongue hanging far out of his mouth as his breath steamed up the window. She smiled and waved back.

Her nerves a bit steadier, she looked back up at the outside of the Beetle above her. It wasn't a very long climb, really. Holding the strut and standing on tiptoes she could almost reach along one outsloping edge to a vertex where the upper face sloped back inward. She readied herself. With an act of will she put aside the drop into nothingness below her and hoisted herself up onto the strut to a handhold on the open shutters of the upper window face.

She caught it, but as she tried to force her other arm up into the same position and get a foot onto the strut to push herself up, everything went wrong. Her foot slipped on the strut, and instead of holding on, her instincts were to move her hands down to break a fall. She fell sideways in slow motion. Bosco's head followed her as she fell and she saw his mouth open and heard a squeal-bark over her radio.

Noah yelled, "Be careful!"

Why do people always warn you to be careful after something bad has already happened? She pondered this as she fell in slow motion past the base plate.

"She fell! I can't see her!" Noah's panicked voice sounded tinny and shrill through her helmet-mounted radio.

All she could see was the Moon in front of her as she hung in space. This was going to be the longest fall in history. She would have a very long time to get used to the idea that she was dead. Likely not enough time. But definitely a very long time.

Then she barely saw something between her and the moonscape in front of her. If there was ever a time for grasping at threads, this was it. And a thread was basically what she had. She caught hold of a long thin line before even realizing what it was: Bosco's leash!

Bosco had been tied to the Beetle's landing gear. His leash was still wrapped through its own handle around the leg above, while the end with a spring clip on it, where it would attach to his collar, dangled below. Dana grabbed the clip and forced her arm down to her beltline. She clipped it to where she had sewn some loops for holding tools.

"Dana." It was Dr. Cavor's voice. "I am going to turn this thing around and accelerate towards the Moon instead of braking. We can get to you. It will throw off my calculations, but I think perhaps we can stop pulling on Earth for a bit and get enough side velocity relative to the Moon for an orbit rather than cratering. If I can work it out, perhaps we can manage a slingshot back, rather than the land and push off again I was originally planning."

Dana said, "Wait. I think I'm ok. If anyone was praying for a miracle safety tether to appear out of nowhere, your prayer was answered. Bosco's retractable dog leash is still out here attached to the landing gear. I barely managed to grab it."

"A miracle indeed," said Dr. Cavor, relief in her voice. "If you figure out which deity or pantheon was responsible, do let me know. I will gladly do a testimonial."

After what seemed like an improbably long amount of time, the leash reached the end of the retracting line. There was a jerk, and Dana bounced around a bit. She saw something red and some flecks of blue flash past her face, and then she was hanging, once again, above the Moon, looking down. But the reorientation seemed surreal and somehow she no longer felt frightened. This was a truly amazing view that quite probably no one else in history had seen. It was a real adventure, and she was the hero!

"Dana! Are you all right?" Dr. Cavor asked.

"I'm still here," Dana said. "I'm ok. But I'm almost literally hanging by a thread here."

"Can you get back up?" Noah asked.

"I only feel a little bit of pull, but is there any chance you could cut the gravity for a few seconds?" She wiggled back and forth until she could get her hands on the leash.

There was a pause, then Dr. Cavor said, "A few seconds should not affect our course greatly. There." Dana felt the leash go slack as she became entirely weightless again.

Dana pulled herself hand over hand, and went flying up past Bosco again, earning yet another excited squeal-bark and more ferocious tail-wagging. She managed to grab onto the shutters of an upper window, and in no time had worked herself up to the top of the spacecraft, into the glaring sunlight. Keeping her helmet faceplate turned away from the sun, she oriented herself until she was flat against the window at the top of the structure.

"I'm here. Gravity on!" she yelled. When the slight sensation of having weight returned, she was laying face down on her belly, on the shutters at the top of the Beetle, looking in at Dr. Cavor's and Noah's smiling faces.

Dana smiled and waved, then went to retrieve her wire strippers and wire nuts from her pouch. She found it empty. She immediately remembered the colors she had seen flash by briefly when she had come to the end of her rope.

"Um, bad news," she said. "I dropped my tools. I mean, normally I would strip the wires with my teeth and twist them together. But that isn't going to work from inside my parade balloon suit."

"Can you close the shutters manually?" asked Dr. Cavor. "It is really getting quite hot in here now." Dana could also already feel her suit heating up from being in the direct sunlight, and the shutters she lay against were so warm she suddenly worried about burning a hole in her spacesuit.

"Hmmm... Yes. I'll need to detach the springs. These gloves will make it tough, but let's see... Yes! That one came loose with no problem."

She maneuvered about, doing the whole window. "Which ones are the other two that are stuck open?"

From inside the spacecraft, Dr. Cavor pointed out the other two open shuttered faces on the sunny side of their polyhedral projectile, and Dana carefully detached all the springs and pushed them closed.

"Ok, I think they will stay closed." Dana said, "Turn off the gravity again, I'm coming home!"

Floating again, she accidentally pushed off from the window she was against and missed getting any new hand hold. She panicked for a moment, vertigo and visions of floating away into space forever, until she remembered the dog leash. Her hands found it, and she hauled herself back to the airlock. That was twice now she would have been lost in space if Bosco had not come with them and brought his leash.

"Thank dog..." she said under her breath.

In a few minutes, with a little bit more crowbar action required to open the inner airlock door this time, she was back inside.

"Wow, my ears really popped there when you opened the airlock," said Noah. "I got scared for a second that we were losing all our air."

"Hang on," said Dana, unclamping her helmet. She went to the spare air tanks. "There, I'm adding some air from the tanks to compensate for the loss and bring us back up to a reasonable pressure. But we can't do that too many more times before our reserves will be empty."

She felt intense relief at being back inside the little spacecraft that had seemed so flimsy before. *Safety is such a relative thing.* Noah and Dr. Cavor both gave her great big relieved hugs, and Bosco licked her face.

"Well then," said Dr. Cavor. "We are not going to overheat. Now it remains to be seen if I did the calculations correctly, so we do not arrive at the Moon going too quickly."

CHAPTER 19 — TOUCHDOWN!

Despite the recent excitement, Dana was actually feeling bored from the days of travel, trapped in the small biosphere turned spacecraft. "Are we there yet?" she asked.

"In fact, yes. We are nearly there," said Dr. Cavor, looking up from the Beetle brains. "We have mostly matched velocity with the Moon's orbit and are now basically coming in for a landing. We are still on target, and I think our speed is good." She punched some buttons and all the blinds on one side of the Beetle opened up.

The Beetle rotated while the two children and a large brown dog looked down at the moonscape around the pentagonal base plate below them.

"Wow!" the children chorused. Bosco whined softly and Dana patted him on the head.

"We will want to drop all the way to the ground," Dr. Cavor said, "in order to get the maximum acceleration when we push off back towards the Earth. And it will also give us a chance to fix the broken shutter while we are grounded, which we will need for the ride home."

"So will we get to walk on the Moon?" Noah asked.

"Well, my understanding is that there is only one space suit, and Dana needs it to fix the shutter wiring. But I do not know how I can possibly deny you the opportunity for a moonwalk. If you think we have enough reserve air to replace the pressure we will lose, then you can have a turn after Dana is done. So, where would you children suggest we land?"

"Tranquility Base!" Dana said without hesitation. "The site where the first men landed on the Moon!"

"Yeah!" said Noah.

"Mr. Bosco? Your vote?" Dr. Cavor asked the dog, who only cocked his head at her mention of his name. "No opinion? Then the Sea of Tranquility it is."

"Sea?" asked Noah.

"*Mare Tranquillitatis,*" said Dana. "There are several named lunar *mares,* which is Latin for 'sea.' Some of the early astronomers thought they might be looking at water, but they're actually ancient lava flows, possibly the result of asteroid impacts when the Moon's crust was much younger and thinner over a molten core."

"Splendid, Dana!" said Dr. Cavor. "They were indeed seas once; seas of molten rock. And since these lava flows are relatively newer ground, they have fewer craters from meteor impacts. A lunar sea was chosen for the first landing site as a smoother place for the lander to touch down. And likely *Mare Tranquillitatis* was picked because when you are riding a series of controlled explosive burns for a few hundred thousand kilometers to be the first humans to set foot on another world, having a little tranquility in the plan somewhere certainly could not hurt."

"Which one is it?" asked Dana, looking down at the Moon under their feet, and they all followed her gaze.

"Hmmm, there, I believe," said Dr. Cavor, pointing toward one edge of the Moon. "We are going to come down a bit lower first, but then we can drift this way as we descend. Once we are low enough, we can move along the surface from here to here." As she spoke, she was tracing an imaginary line across the air in front of her, between them and their view of the Moon, as if following a road on a map.[75]

"So we're going almost to the dark side, there?" said Noah, pointing. "That's the side you can't see from Earth and that we don't have maps of, right?"

---

[75] People say, "the map is not the territory," but in this case, the map she used literally was the territory. Or maybe it's called *lunitory* on the Moon? *Terra* is Latin for "Earth." *Luna* is Latin for "moon."

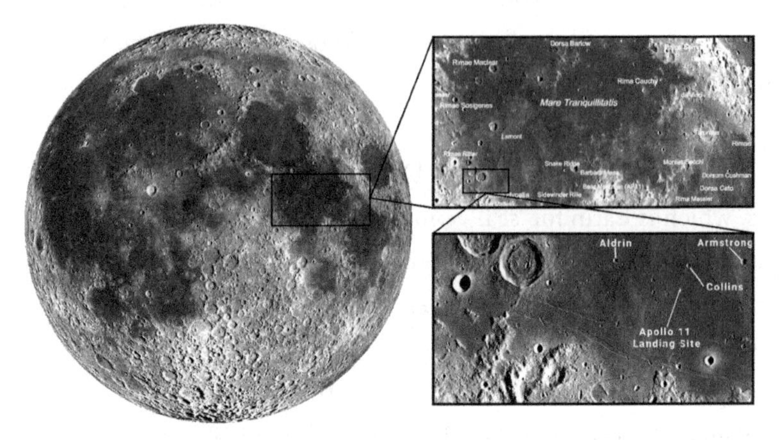

Images from NASA LROC https://quickmap.lroc.asu.edu/

Dana laughed and began to correct him, but seeing the interested look on his face, it occurred to her that he took being corrected about science facts a lot better than she did.

"There's the dark side and the far side. They're not the same thing. The dark side is the side away from the sun." She stopped and thought, then said, "Although the Moon is pretty full, so I guess they're pretty much the same thing right now."

"Ah, not quite," Dr. Cavor now corrected Dana. "We have been travelling for several days, and the Moon has looked quite full to us the whole time. It still does. But on Earth it would be visibly waning. We are no longer seeing it from the same direction as someone standing on the Earth. So the side there, opposite our destination, is a bit of fully-lit far-side territory we can see. The other seven billion human beings in existence cannot see that part of the Moon, and, while we do have maps of it thanks to lunar satellite mapping missions, those maps have certainly not existed very long. You are looking at a piece of the moon that almost no other human beings have seen, with their own eyes, in all of history."

Dana was awestruck, and a reverent silence followed Dr. Cavor's words. At least, until Noah broke it.

"Well, it looks the same as the front side," he said.

Dana sighed. He liked bugs and bacteria. Space stuff didn't have the same appeal. Still, how could he not be awed by where they were and what they were seeing?

"Actually," Dr. Cavor began to lecture again, "the far side looks somewhat different. It does not have the same darker lunar *mares* we see on the earth-facing side. The theory is that the lunar crust on the far side cooled faster and thicker because it was facing away from the heat of the still-molten Earth, so meteors did not pack the same punch and make as many lava-seeping holes on that side."

Dr. Cavor continued to talk about the Moon, while Dana stared down at it. Noah lost interest and went to check on the cyan-aid.

When they were low enough to see the lunar horizon from only the upper windows, Dr. Cavor partially opened the shutters at the top of the airlock. This let the Moon and the neg-grav of the cyan-aid tanks "see" more of each other, and the feeling of internal gravity increased while their descent slowed. Dr. Cavor continued making adjustments until they were hanging, nearly motionless, over the lunar terrain.[76]

"Now," said Dr. Cavor, "we need to plot a course for the Sea of Tranquility. We will open only one of our five lower-side windows in the direction we want to go, and the moon will pull us in that direction." The Beetle shifted slightly and began to accelerate across the lunar sky.

"Hey," said Dana, "any chance you'll fill us in on the whole negative-gravity thing? We've been playing with that machine from your lab for a while, but we haven't come up with any decent ideas about how it actually works."

"I told her to accept it as magic," said Noah, "but she really wants a scientific explanation. It's been bugging her all along."

[76] No, this book will decidedly not be using the word "lunain."

"Ah," said Dr. Cavor, "well, any sufficiently advanced technology is indistinguishable from magic."[77]

"Is the explanation really that far beyond us?" asked Dana.

"Perhaps not," admitted Dr. Cavor. "Have you heard of things like quantum spin, and theories of eleven-dimensional reality?"

"Sure," said Dana, "But I'm not going to claim that I understand that stuff."

"Commendable, to know one's limits," said Dr. Cavor. "You do know what a dimension is, certainly?"

"Sure," said Noah. "You can measure any position with three numbers, height, width, and depth, and time is sort of a fourth dimension."

"Yes!" said Dr. Cavor, "So three dimensions in which we may move freely, and time in which we seem to be constrained to go only one way. And if there are seven other dimensions, or even seven million, they are not as obvious."

"Ok," said Dana, "but how does that get us to neg-grav?"

"Well, one useful model for how reality works is that all of the forces that act across time, to move things around in the other three obvious dimensions, are controlled by connections, movements, or vibration in other dimensions. But there are dimensions too small to be immediately obvious to us. Hmm... how can I best describe this?"

She paused and her forehead crinkled in thought. She looked down at Bosco, who had his head at the window, tongue out, watching the moonscape rushing past, kilometers below them. Her glasses slid down her nose. She pushed them back into place and looked up at the children again.

"If we were to keep going in one direction forever, at a speed much faster than light, it is likely that we would eventually end up in the same place again — that what appears to be an infinite distance dimension

---

[77] A quote from Arthur C. Clarke (1917–2008), author and futurist. Possibly the first to imagine placing communications satellites at geostationary orbital distance, sometimes called Clarke Orbit or the Clarke Belt.

is really a closed loop. But the loop has expanded to be so large that we cannot ever see that it is a loop.

"But we can also imagine dimensions so small that, heading one direction or the other, we are back in the same place we started so quickly we do not even notice that we have moved. We could be traveling in these collapsed dimensions all the time and never notice, except that such travel might change the way we interact with other things in the three obvious dimensions. Such motions might control things like gravity, electromagnetism, nuclear forces, and so on."

"Ok, I can kind of imagine that," said Dana. "Although I'm not sure how forces happen based on that." Noah nodded.

"It is certainly not intuitive," said Dr. Cavor, "but there is math that seems to work based on these ideas. Now, as for how we get from that to negative gravity... do you know what a wormhole is?"

"I've heard of that," said Dana. "Like a way to jump from one part of the universe to another, faster than light. A hyperspace gateway, like in science fiction."

"Yes," said Dr. Cavor. "But it does not have to go that far. Imagine a wormhole a few decimeters[78] long. Like from one end of that machine, which you 'borrowed' from my lab, to the other."

"But what's the point of that?" asked Noah.

"The point of a wormhole is that it can break out of our familiar three dimensions, and then reattach back to them," said Dr. Cavor. "Well, it turns out that there is no reason why it cannot twist things around a bit along the way. Twist it properly and things that were spinning one way going into a wormhole can be spinning the other way when they come out. And if the spinning is reversed in the correct tiny dimensions, we can select for different properties on the other end. When it gets back to our boring normal three dimensions, the thing that made the trip can interact very differently with material that did not make that same trip."

---

[78] A *decimeter* is one-tenth of a meter, about four inches.

"So, reverse the spin in some tiny gravity-controlling dimension, and you get neg-grav?" asked Dana.

"That's the essence, yes."

"What else can you reverse?" asked Noah.

"Oh, very nearly any basic physical property," said Dr. Cavor. "Although we do need to invert dimensional orientation in pairs. You cannot only do one. Which is why my machine is set to flip right and left as well as gravity. But if you wanted, you could reverse every physical property and make true antimatter."

"I've heard of antimatter," said Dana. "Isn't it super unstable and will immediately explode?"

"Not unstable on its own. And the antimatter you are thinking of is not reversed in all dimensions; it is simply matter with the reverse of the usual electromagnetic charges. If it comes into contact with normal matter, because it has opposite charges, things become interesting. Particles and antiparticles annihilate each other, with all the mass converted directly into energy in an extremely efficient explosion.

"When you create that kind of antimatter, you have to store it in a vacuum because it would react with normal air. One gram of electromagnetically reversed matter, annihilating with one gram of normal matter, would release more energy than both of the atomic bombs the United States dropped on Japanese cities in World War II combined."

"See, Dana?" said Noah. "I told you not to mess with the settings on that machine."

"Oh, yes." Dr. Cavor smiled oddly. "That could have been very bad."

Dana's face paled a bit as she bent down to scratch Bosco and look out at the lunar landscape moving below her.

After a number of hours of travelling to the lunar east, the sun was no longer directly above their heads, the way it had been for much of their trip, so they were able to open the shutters on two of the forward-facing upper windows without overheating.

"How long until night?" asked Noah.

Dr. Cavor answered him. "If we keep traveling at this rate, we would be around to the dark side quite soon. But we are close to our destination now. When we stop, night will come slowly. As you can see, looking at the ground below us, the shadows are already very long, but the moon rotates very slowly. The lunar day is nearly a month long, and full night will not be on us for well over another full Earth day."

"How close are we?" asked Dana.

"We are already over the Sea of Tranquility. Now we are looking for the *Apollo 11* landing site." Dr. Cavor was peering out the window, scanning the horizon. "And, perhaps... Yes! I think I see what we are looking for."

She fiddled with the shutter controls, opening shutters on two lower windows behind them, one at a time, to put on the brakes. The ship began to slow down. Dana and Noah crowded the front-facing windows, looking for what Dr. Cavor might have seen, but it was quite a few minutes before anything became visible to them. Dr. Cavor must have really good eyesight, even with those glasses.

Finally, Noah said, "Over there!" and Dana looked where he indicated. There was definitely something there reflecting sunlight back at them; it glinted near the horizon, getting brighter as they approached. In what seemed like no time at all, they were descending slowly towards the *Apollo 11* lunar landing vehicle — the site of the historic first moon landing.

"What happened to the lunar lander?" asked Noah. "It looks like someone cut the top off."

"That's the descent module," said Dana. "The top part of it took off again, with the astronauts in it, to link up with the orbiting part of the rocket and head back to Earth."

"Oh, right," said Noah. "I knew they went home. I should've been surprised that there was anything still here at all."

"The *Apollo* vehicles broke into a lot of pieces along the way," Dana said. "The original Saturn V rocket was gigantic. Sections fell away during the launch as fuel was used up, then again as they left Earth orbit and headed for the Moon. The lander went down while one of the astronauts stayed in the orbiter above. Then only part of the lander came back up. They even left a last section behind right before re-entry into Earth's atmosphere."

"Our spaceship is a lot less wasteful," said Noah.

Dr. Cavor interjected, "Solar scientists are starting to develop reusable rockets now, but shedding sections as they went was state-of-the-art for the 1960s, and they made it work well."

"Solar scientists?" asked Noah.

"As opposed to our Lunar Society," answered Dr. Cavor. "Solars is what we Lunars sometimes call 'regular' people, those who do not have access to the secrets of the Lunar Society. It is mostly a label to keep track of which sphere created tech and who uses it. Solar, Lunar, and Xeno."

"Xeno? Like aliens?" asked Dana.

"Yes," said Dr. Cavor, "Well, perhaps. Although there is certainly evidence that they have been around on Earth longer than human beings have — wherever they might have originated."

Dana's head was reeling. *We are not alone in the universe, and a school principal knows all about it.* She was about to launch into a slew of questions when there was a bump.

Dr. Cavor said, "Contact. All stop," then did something to the controls. Shutters on the lee side of the sun opened.

Dana looked around her, then said excitedly, "Tranquility Base! We're here! The Beetle has landed!"

Dana felt excitement building inside her as she remembered that she was going to put on her spacesuit again. She was about to walk on the Moon!

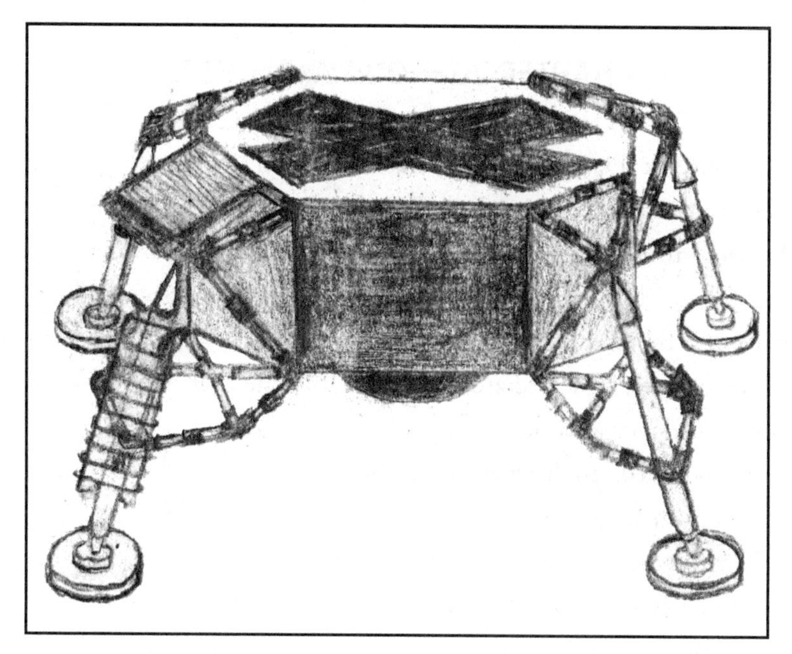

From the notebook of Dana Dash

CHAPTER 20 — ONE SMALL STEP

In the airlock, Dana rechecked that she had all the materials she would need to finish the repairs she had first attempted during her spacewalk, but her mind wasn't really on that task. She was going to step out onto the surface of the Moon! She would be the youngest person, and the first female, and the first person in fifty years. As she closed the inner airlock door behind her, Bosco scratched at it to indicate that he too would like to step outside for a while.

Noah pulled him back and said, "Sorry boy. If we had extra space suits for either humans or canines, we'd all go for a moonwalk."

Dana turned on her radio and attached her helmet securely, wedged the crowbar into the outer airlock door in the floor, and said, "Ok, here we go." She pushed down on the crowbar, slowly letting air escape from the airlock as her spacesuit became rigid, proving itself once again airtight. She then lowered herself through the door in the floor onto the landing plate below.

She narrated her movements over the radio for the others still inside, "I'm on the landing plate. Making my way over to the edge. The plate is only depressed in the surface about five centimeters. The surface is like powder... I'm going to step off the landing plate now."

She stepped off the landing plate, onto the lunar surface, saying loudly and clearly for her audience, "That's one small step for [a][79] girl, one giant leap for girlkind!"

The cheering of Noah and Dr. Cavor from inside the spacecraft came over the radio, as well as a couple of excited squeal-barks from Bosco. "The surface is powdery. I can pick it up with my toe. It sticks like powdered

[79] There's some controversy over whether or not she actually said the word "a."

sugar to my boots. I only go in maybe a couple centimeters, but I can see my footprints."

Then she spotted other footprints. Those of the men who had been here before her. Fifty years ago, they had walked right here where she was walking now. With no atmosphere, there was no wind, no rain. The footprints were perfectly preserved.

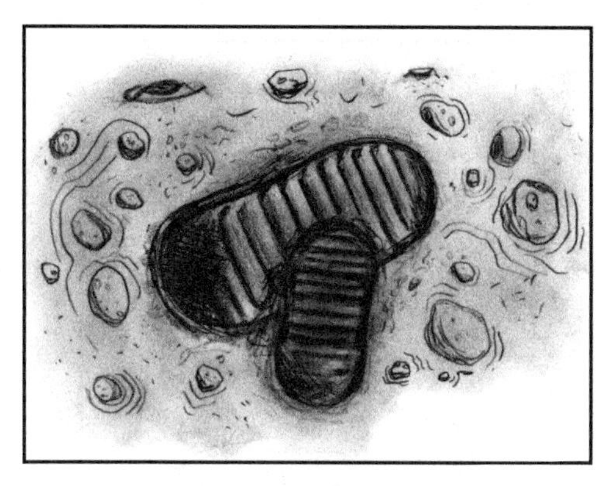

From the notebook of Dana Dash

Carefully she lowered her right foot into the middle of one of the older footprints, then stepped to the next, heading towards the abandoned lunar landing engine stage of *Apollo 11.* She marveled that she was literally walking in the footsteps of one of the first two people to ever set foot on the Moon. "Dana," Dr. Cavor's voice came through the radio. "First things first. You must do your homework before you play. The repairs await. I am disconnecting the power to the shutters now."

Dana sighed. "Yes Dr. Cavor." She obediently bounded up the side of the Beetle like a low-gravity mountain goat and made quick work of the repairs. She jumped off the Beetle and landed softly on the Moon's regolith[80] and resumed her trek to the *Apollo* lander.

"It's fun walking out here in one-sixth gravity," said Dana, "I'm getting used to it. It's kind of a hop and glide. Talk about having a bounce

---

[80] *Regolith* is the proper term for material found on the surface of the moon.

in your step!" Noah and Dr. Cavor both laughed.

She reached the corner of the *Apollo* lander that had a ladder on it and put her hand on one of the rungs. Even with the sun very low in the lunar sky behind her, through her space suit gloves, the ladder rungs were quite hot to the touch. Between the third and fourth rung from the bottom, she found an engraved plaque, curved to fit around the landing gear leg, to which the ladder was attached.

From the notebook of Dana Dash

She read the engraved inscription aloud: "*Here men from the planet Earth first set foot upon the Moon, July 1969 A.D.* Hey! The astronauts all signed this thing. Neil A. Armstrong, Michael Collins, Edwin E. Aldrin. Did you know that Buzz Aldrin's first name was really Edwin?" Noah admitted that he had not known that. "And Richard Nixon got to put his name on here too," said Dana, "even though he never got anywhere near the Moon. Just because he was president."

She smiled to herself, then reached into her belt pouch and pulled out a Sharpie permanent marker. It was a bit messy, as residual pressure inside the marker had forced some ink out of the tip, but she managed to tag the plaque with her name and a quick doodle before recapping it and returning it to the pouch.

Dana noticed some color on the ground about ten meters off to her left and bounced over to check it out.

"The flag fell over," she said. She picked it up. The side that had been facing up was somewhat faded from the sun, but probably not nearly as much as it would have been in an oxygen atmosphere. "Aren't we supposed to burn it or something because it touched the ground?"

"I'm gonna say that touching the Moon doesn't count for that," said Noah.

"Cool," said Dana, brushing the regolith off. Then she set it back up to stand proudly. Her father was always ranting about the terrible things that the government did, but she figured that, despite any of that, the very first Moon landing had to be something you should be able to feel a little national pride about.

"Ok, my turn!" said Noah, "When I get out there I want to lay down and make the very first moon-dust-angel ever!"

"Ha," said Dana, "not if I do it first, right now!" She started to lower herself to the ground.

"Hey! No fair!" said Noah, "That was my idea!"

"Please be nice Dana. Now come back and give Noah a turn moon-walking," said Dr. Cavor. "I do not recommend that anyone lies down. The ground out there is still quite hot, those moon rocks look like they are sharper than rocks that get weathered, and your homemade space suit is not very thick."

"I was kidding," said Dana, getting back to her feet. "One more thing before I come back." Dana bounded back over to the *Apollo 11* lander. With one last super-bounce, she leaped up the ladder and climbed on top of the lander. "Noah, take my picture standing up here."

"What's the magic word?" said Noah.

"Abracadabra? Alakazam? Opensesame?" she teased.

"Try again," said Noah.

"Ok, pleeeaaase will you take my..." She trailed off as something caught her eye.

She turned towards the west, shielding her eyes against the sun with her hand. Something was moving.

Shielding her eyes against the blazing sun, Dana peered out at the vast landscape. Deep craters and long black shadows made a black and white patchwork quilt of the Moon's surface. Could she really have seen something move?

"Dr. Cavor... Noah... I'm not sure I'm alone." But they *had to be* alone — the Moon was unsuitable for life. There couldn't be anything out here except them, their spacecraft, and the fifty-year-old evidence of the first Moon landing. Looking back at the Beetle, she could see Noah, brow furrowed, squinting against the sunlight, his nose smushed up against the window, his breath fogging it up.

"She's right," he said to Dr. Cavor. "There's something moving out there."

Dr. Cavor joined Noah at the window. "Where?"

Noah pointed.

It was no longer nebulous motion on the horizon. Dana could see small, distinct shapes scuttling out of a crater, headed towards them. She wasn't sure if they looked small because of the distance or if they really were tiny. The craters, which were the only landmarks, came in all sizes, making distance hard to judge.

"This is impossible!" Dana exclaimed. "There *can't* be any life out here. There's no atmosphere!"

"I know," Noah said. "That's why there are such major temperature changes between night and day. It can get up to a hundred degrees celsius and as low as negative one hundred and seventy degrees celsius!"

"Exactly! That's why this can't be happening." She looked back out at the fuzzy life forms getting larger. "Bugs? Are they alien bugs?"

"They're too big," Noah pointed out.

He was right. Now that they were closer, Dana could tell that they were surprisingly large. And closing the distance surprisingly fast. They were over a meter tall and they moved in short bursts on all fours. Wait, no, did they have six or eight legs? Eight legs. *So some sort of moon-spider? A lunar-tick?* They rose up on their back two or four legs to leap-bounce over the moon's surface. Dana counted at least seven.

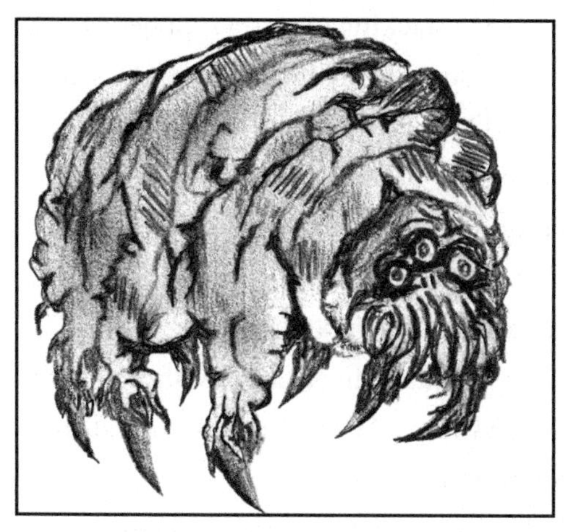

From the notebook of Noah Knight

Their legs were not long, and ended in hand-like claws, but out of the center of the palm of each hand protruded a long spike. The spike came out whenever a limb touched the ground, piercing the lunar soil. So instead of any sort of footprints, the left tiny round holes in the dust. They were close enough now that Dana could see they had three forward-facing eyes. Didn't forward-facing eyes mean predator? But she was on the Moon — who knew what biological rules applied to moon predators.

"Dr. Cavor, what should we do!" Noah's voice shook.

She smiled thoughtfully at the challenge. "Dana is still outside, and I detached the wiring that runs the shutters, so we cannot take off to escape them. This really is a predicament." Bosco whined at Dr. Cavor's words. "I suggest we stay put for now; it is doubtful they will be able to break through the polycarbonate walls. Dana, get down and try not to be seen. Noah, be ready to barricade the airlock. I will see if we have anything on board that can be used as a weapon, if it comes to that. Where is that crowbar?"

"It's in the airlock, so I can get back in," said Dana as she crouched down on top of the *Apollo 11* descent stage. There was a hole in the top middle, where the upper section had been connected, and it was big enough for her to squeeze into. Her head peeked out as she watched the moon bugs approach the Beetle.

She could see Noah stacking up extra supplies against the inner airlock door, but Dana highly doubted that would stop the space bugs if they made it inside the airlock. Dr. Cavor seemed surprisingly calm as she observed the bugs surrounding the little spaceship.

"Noah, what do you think they are?" Dana asked.

Noah thought about it for a few seconds. She had finally abandoned expectations of an answer when he did speak.

"They'd have to have amazing heat resistance to withstand the daytime temperatures, and possibly a layer of fat like a seal to keep out the cold. But that alone might not be enough..."

"They could be similar to tardigrades," Dana suggested.

Noah brightened up immediately at the mention of his favorite subject. "That's true," he agreed. "They can live in a vacuum. Did you hear about the tardigrade space experiment?"

"Of course. Scientists sent them into space riding the FOTON-M3 to analyze their behavior!"

"They were exposed to insanely extreme temperatures and even the vacuum of space but still survived!" Noah exclaimed. "The females even laid eggs while still on the FOTON. It is possible of course that these things," he gestured towards the alien bugs, "have similar traits to a tardigrade, but it's doubtful they're related."

"Why?" Dana asked. She hated seeming oblivious to his reasoning but curiosity got the better of her.

"Because, while tardigrades come in many sizes and species, they're all microscopic. Besides, even if these moon bugs do actually look a bit like tardigrades, and may have at least some similar characteristics, we should consider that tardigrades essentially hibernate when conditions aren't livable and these creatures look *pretty* active right now."

Dr. Cavor interrupted their scholarly discussion. "Take care! Remember what I said about geniuses stopping to think about the reasons for the orange stripes while the tiger is charging them?"

As if to emphasize Dr. Cavor's words, one of the space bugs crashed into the Beetle, smashing against the transparent polycarbonate windows. Even Dana, who was trying to remain hidden, many meters away, flinched at the sound over her radio. Bosco growled and began barking frantically.

Instead of trying to break in, the bugs surrounded the Beetle, and working with surprising speed and intelligence, pushed against it as one, lifting it off the lunar surface. Then they began carrying it back the way they had come.

Dr. Cavor sighed. "Dana, it looks as if we are going to be taken for a ride. Do you think you can follow along without being spotted? I am going to try to get the power reconnected so I can launch. Then we can circle back and pick you up."

"I'll do my best," she answered.

The moon bugs were moving amazingly quickly, carrying the Beetle along with them. Dana waited until they were far enough away that she could risk leaving her hiding place. She jumped down off the fifty-year-old lunar lander and started to hop-jog after her own retreating spacecraft — the one that represented her only chance of ever getting home. The inflexibility of her homemade space suit joints made her attempt at rapid motion across the lunar surface somewhat comical. It was also hard to see the retreating spacecraft. They were heading west into the setting sun and Dana could not easily look in that direction without risking permanent eye damage. So she kept her head down and her body bent forward, looking up only occasionally from the ground right in front of her to try and catch a glimpse through squinted eyes. She could only imagine how she looked, bouncing along with her arms held straight out to the sides, looking at her feet, like she was trying to balance while hopping on a high wire.

She had given the moon bug kidnappers quite a bit of a lead to begin with, and despite having to carry the Beetle with them, they were getting farther away every second. Fortunately they didn't go so very far. It was only maybe a minute or two later that Dr. Cavor spoke again to Dana over the radio, still sounding surprisingly calm.

"How is your supply of oxygen, Dana?" she asked.

Dana checked the number on her pressure gauge. "I'm at about fifty percent."

"We have stopped at some sort of opening in the ground. I think they are taking us down under..." Up ahead, the Beetle disappeared from view. Line of sight lost, all Dana could hear over her radio was static.

Dana stood there, all alone on the surface of the Moon, in her poorly designed homemade space suit, her air tank half empty.

CHAPTER 21 — DOWN THE HATCH

She looked up at the at the crescent Earth hanging in the sky above her. "Half full!" she said out loud to herself. "My tank is half full. Must try to remain optimistic..."

Dana kept moving until she reached the spot where the moon bugs had disappeared along with her companions and their spacecraft. It looked like another of many similar sized craters on the lunar landscape, but when she bounced down into it, the floor of the crater felt more solid under her booted feet. She realized that it was missing the several centimeters of moon dust that covered the ground she had been walking on before and she was no longer leaving footprints.

Looking more carefully, she saw a line running across the bottom of the five-meter crater, dividing it exactly in half. She realized that she was actually standing on a pair of large doors — some sort of round hatchway — designed to look like the moon's surface. The doors were closed now but would have been easily large enough to swallow the Beetle when open.

After failing to get her fingers into the crack, and wishing she had brought the crowbar outside with her, Dana started looking around the crater for some sort of lever or switch — anything which might open the doors so she could follow after her friends. She searched the floor first, then tried all around the maybe two-meter high rim. *I have to get inside! Who knows what those moon bugs eat — hopefully not people or dogs. But nothing good can be happening to them down there. I'd better figure something out soon if I'm gonna have any hope of rescuing them.*

On her second circuit of the top of the crater rim, coming back to where she had started, she looked at the footprints she had left on her bouncing jog from the *Apollo 11* lander. Then she noticed that, while difficult to make out, the footprints of the moon bugs were not entirely invisible.

The small holes their spiky leg extensions left in the dust were barely visible, but where a large number of them had passed, there was a dimpled effect that was different than undisturbed moon dust. Moving around the top of the crater again, she found a couple of other directions in which the dust had been similarly disturbed and made up her mind. Picking one at random, she set off again across the lunar surface. She hoped that she would find some more accessible door into what, she assumed, would be an underground complex.

On the positive side, the new direction she was traveling was not directly into the setting sun. She was moving roughly north by north west and only had to keep her head turned a little bit to the right to keep the glare out of her eyes. Not having to keep her head down made it easier to establish a more regular skipping gait across the lunar surface. She tried to estimate her speed, guessing the distance and duration of each of her hops. As her pace improved, she decided that, despite the seeming awkwardness of this mode of travel, she was probably moving at over twice the speed she could have run on Earth. And it was amazingly effortless. Her Earth-grown limbs had amazing strength and stamina in the one-sixth gravity.

On the minus side, however, even though her space suit was white and thus fairly light-reflective, the direct rays of the setting sun, with almost zero atmosphere to soften them, made the inside of her space suit really hot.

The shadows cast by rocks and hills were amazingly black. It was hard to see the bottom to judge the depth of even small shallow craters, and when, out of curiosity, she tested the temperature of the ground in these long lunar evening shadows, it was almost as painfully cold as the lighted areas of ground were still painfully hot. When the heat became unbearable, she was able to rest and cool quickly by sitting or lying down in these shadows, letting the cold rock suck the excess heat from her body. But time wasted on cooling meant air spent with no additional distance gained. She didn't know if she would end up passing out from heat or lack of oxygen before she found out where this bug trail led.

As she moved, she tried to distract herself from both the physical discomfort and the mental agitation that came with any thought of her current situation or that of her friends taken with the Beetle. She counted rocks and craters. She tried to remember what she knew of the lunar geography of this area. She pictured a map she had seen of the *Apollo 11* landing site. There had been a large crater, more than five kilometers wide, south of the landing site called Moltke, if she was remembering it right. But she was heading north. She remembered three smaller craters in a west to east line in that direction. They had been named after the three *Apollo 11* astronauts: Aldrin, Armstrong, and Collins. She would be heading towards the middle one. She could see a bit of ridge ahead of the otherwise flat landscape and thought it was probably one of those three craters. But she couldn't decide which one that was.

She began to obsess on the question, trying to picture the map in her mind's eye. She bounced on for what seemed like hours but was probably only a lot of minutes. She resisted checking the oxygen gauge. It didn't really matter what it said at this point; either she would make it somewhere in time, or she wouldn't. *Was it Collins or Aldrin in the middle? I'm pretty sure Armstrong was the one on the right of the map — to the east.*

Then she fell for the first time. Her foot came down in a small crater covered in the dust and caught the edge as she launched off again. She tumbled through the air, head over heels, and her gymnastic training, which might have worked on Earth, failed her in the low gravity. She landed on her rear end hard and bounced into the air again, spinning faster. She hit the ground again on her head, but with her hands up to cushion the blow, and rolled for quite a way before coming to a stop.

Get yourself up! Must keep moving!

She glanced at the oxygen gauge again, despite willing herself not to. Less than 25%. She fell three more times, and each time she allowed herself a peek at the gauge, until she was in the red zone and she knew that she was probably not going to make it. But the ridge was much closer, taking up most of her forward view now. Then she realized she was going uphill.

There was less dust now and more bare rock, but she could still see the dimples created by moon bug legs in the pools of dust where it collected in ocasional depressions.

It was a last push of her physical ability to make it to the top of the ridge. She was moving only slowly, probably quite dehydrated, most of the water she'd drunk having been sweated out to collect in her boots. Her air gauge was down to 2% when she finally went to her knees. She managed to crawl the last few meters to the edge of what, she was now almost sure, was the crater named after astronaut Michael Collins. She managed to place herself in the shadow of a large rock on the crater rim and sighed in relief as the cold moon rock pulled the heat out of her body.

The crater below her was alive with moon bug activity — or at least part of it was. The very slowly setting sun still lit up the top part of the eastern rim. Moon bugs, looking somewhat different than the ones that had taken the Beetle earlier, were busy harvesting a crop growing on the crater wall. Teams of these moon bugs moved along the line of light and darkness, cutting something green and fungus-like from the stone with blade-shaped forearms like a praying mantis. Behind each team of cutters, a large harvester vehicle of some sort moved along at the same pace, scooping the cut substance up into a mouth-like opening. As she watched, one of the harvesters turned down the cliff, while another came up and took its place. She saw that there were several lines of these vehicles coming up from the blackness at the center of the crater and returning there with a full cargo. They must be going down into what was, hopefully, the same sublunarian tunnel complex where her spaceship and friends had been taken.

With her oxygen almost gone, she didn't have a lot of time to make a plan. So when an idea occurred to her, she didn't care how terrible an idea it probably was. It was a chance. And any chance was better than what she had thought she had a few minutes ago.

She moved around the top of the crater until she thought she was directly above one of the moon bug harvesting teams. The fungus grew up from trunks, sprouting from the rock like giant lumpy mushrooms. The trunks higher up the crater wall bent almost 90 degrees, so that all the mushroom caps pointed almost directly upward. The tops of the caps were mottled light and dark gray colors that made them blend into the lunar landscape from above, but underneath everything was a bright green.

From the notebook of Dana Dash

Her idea was that, if she could make her way down, unseen, between the fungus stalks, she could sneak aboard one of the large harvester vehicles and it would take her down into the tunnels where she might hopefully find her friends and more air — not necessarily in that order of importance to her right now.

She knew she had miscalculated almost as soon as she went over the edge of the crater. The thick green stalks had looked solid enough, but did little to support her weight — instead, they crumbled and dissolved into sticky green goop as she tried to grip them. At first she thought that, even in the low gravity, she was going to tumble all the way down the cliff, failing to get a handhold as she bounced between the stalks like a pachinko ball, her spacesuit being covered with the sticky green substance as she went. Finally she managed to connect directly with one of the thick stalks. The outer part of it exploded away, but there was a strong thin inner core, and it stopped her.

She clung to the stalk for a minute before managing to pull her face out of the goop and wipe off her faceplate to assess her situation. She realized that she had only managed to stop at what was almost the very last row of remaining unharvested fungus. Then, looking to her right, she realized that the harvester bugs were almost upon her. They were moving faster than she had been able to judge from the top of the cliff, and she was too tired to think or act. Before she had time to do more than groan in despair, one of the bugs was at her mushroom and both of its deadly-looking scythe forearms flashed out towards her.

The world tilted as she fell. She realized she was still clinging to the trunk of the mushroom and it had come loose from the rock. The two scythes had passed above her and below her, likely only missing her by centimeters. She rolled as she fell, the view spinning around her, barely able to see where she was heading.

The harvesting vehicle, which she had planned to try to sneak a ride on, was following behind and below the scythe bugs, catching the falling cut pieces of mushroom. Falling towards it, she realized with horror that what she has taken for a large vehicle was actually another type of moon bug. It had a body like a huge millipede with an uncountable number of legs. What she had previously thought of as a mouth-like opening was really an open mouth, into which the mushroom pieces were being scooped by specialized front limbs. Dana let out a scream inside her space helmet as the giant worm-like bug swallowed her whole.

The last thing she saw, before everything went dark, was that the giant bug's mouth contained row upon row of very large shiny white teeth.

CHAPTER 22 — WASTE COLLECTOR

When the giant moon worm swallowed Dana whole, she was surprised to find that its teeth were relatively soft — something more like giant fingernails than teeth, really. They didn't penetrate the space suit but moved in a coordinated rhythm to push Dana, along with the many chunks of green fungus, farther down the worm's throat. Dana was buried alive in the goopy green mess and felt pressure from all directions as more and more fungus was pushed in on top of her. All light was cut off and Dana, while not normally claustrophobic, was about as terrified as she could be. "Swallowed by a giant space worm and suffocating in its gut as it slowly digested you" had not even been on her radar as a possible way to die. Yet here she was.

But after a minute or two of sheer panic, she realized that she was not actually suffocating. She was packed tightly in warm fungus, her arms pinned at her side in the pitch blackness, but she certainly wasn't dying quickly. In fact, she was barely bruised. The pressure on her was noticeable, but her balloon suit pushed back, and it was only a little bit uncomfortable. The oxygen tank on her back was almost out of air, but not quite, and she could still draw some air from the regulator.

At first she could feel the changes of pressure as the worm continued to swallow more of the harvested fungus. But after only a few minutes that stopped and she only felt a sort of constant throbbing. Maybe it was the worm's heartbeat or the constant motion of its many legs? Cocooned in this warm dark place, with very little gravity, Dana started to feel surprisingly relaxed and a little bit of hope managed to creep in to mix with her fear. *Maybe whatever chemical process the worm uses to digest moon mushroom won't work on my spacesuit. The worm might get a bellyache and have to spit me back up. The question is, will my air last?* She employed the oxygen-conserving techniques that Dr. Cavor had shown her aboard the Beetle, and her breathing slowed as she dropped into a meditative state.

She had no idea how much time had passed when something finally changed. She came out of her meditative state to find that the throbbing sensation had stopped and that she was moving again — being moved around inside the giant worm's long body by regular contractions. She was aware that the air in her helmet was pretty stale, and the regulator no longer seemed to be feeding her a hiss of air on her inhales. Then her helmet was being pushed against something more solid and she stopped moving inside the worm. But the contractions and pressure didn't let up. If anything, they increased, threatening to bend or even break her neck. She did her best to keep her neck straight against the pressure. Just when she thought her neck could take no more, something else gave. Her head popped through what felt like a tight rubber ring.

There was light again and she could see. And what a sight it was. She was suspended above the edge of a long, gigantic trough, full of a lumpy green slurry. She was immobile, her head facing to one side, looking down the length of the trough, to where it disappeared into a wall. Several of the giant moon worm harvesters were backed up to this trough and they were squirting out spurt after spurt of thick goopy green *stuff* into the slowly moving river of gook below.

I am being pooped out of a giant alien bug butt, she thought with some dismay. The moon worm bore down again with an even stronger contraction — one that threatened to crush Dana's shoulders. It didn't work at first but, with the next push, the worm managed to clear the blockage and Dana popped out of the myriapoda's[81] bottom like a champagne cork.

Unfortunately, the oxygen tank on her back didn't make it. It ripped free of her back and the hose ripped free of the regulator built into Dana's

---

[81] Worm-like bugs with many legs, such as centipedes and millipedes, are classified in the subphylum Animalia-Arthropoda-**Myriapoda**. The Latin word *myriapoda* means "lots of legs."

face plate. The regulator also almost ripped free of the helmet, left hanging by only a few centimeters of the adhesive sealant that had held it in place, and this opened her helmet to the outside environment. Then the smell hit her so hard that she didn't even have time to wonder about air pressure or atmospheric gases as she sailed slowly through the air, still in one-sixth gravity. She fell for quite a while and had time to be gagging on the stench before she even landed in the goop below, with a slow-motion splash.

It was only when she struggled back to the surface that she wondered why she wasn't already dead. There was air here, and despite the smell — certainly as bad as anything she had ever smelled before — it didn't seem to be poisoning her. The closest thing she could compare the stink to was the smell of a very ripe soft cheese. As she struggled there in the poop trough, trying to stand up, it seemed like her whole smelly world was made of green cheese.

Movement was actually now somewhat easier with her space suit no longer ballooning with internal pressure. With a little effort, she got to her feet. Kicking and clawing, she managed to propel herself up the side of the trough and grab hold with both hands. She did an effortless chin-up until she was doing a Kilroy[82] over the top, taking in the huge room.

It was a vast underground chamber, cut from moon rock, with a ceiling height of at least ten meters. Light was provided by irregularly spaced glowing orbs hanging down from above. She was between two of the large harvester transport worms, each currently dumping their loads, but she could see past them to a large area. The space was big enough for the huge worms to turn around, with several potential exit tunnels visible. Many other types of moon bugs moved around the arriving and departing transport worms. Dana recognized the harvester scythe bugs that had nearly beheaded her and the first type of moon bug she had seen back at the

[82] A meme dating back to World War II, which was a doodle of a cartoon peeping over a wall, usually accompanied by the graffiti "Kilroy was here." For what it's worth, graffiti vandalism is a misdemeanor.

Apollo 11 landing site — the ones that had taken her spacecraft and her friends. There were also some new types of bugs. Had Noah been here, he would have been mentally studying and cataloging, but Dana's mind was on the business of finding her friends and escaping.

There would be no way she could stealthily reach any of those tunnels. While the harvester bugs had not seemed to even notice her existence, the original bugs — dust runners, she decided to call them —had definitely shown an interest in the human beings. Still, she looked for a while at those tunnels and thought, before finally turning her gaze to where she really didn't want to go. But she knew it was her only real option. She finally looked down at the stinking green river of poop-cheese-fungus-fondue and where it disappeared through the wall into a distressingly low-ceilinged but unguarded tunnel.

"Well, here goes nothing," she said to herself. She released her grip on the edge and dropped slowly back into the trough.

She entered the smelly green muck once more and half waded, half swam towards the dark opening. There was only a little bit of room between the rounded roof of the tunnel and the surface of the sludge and she bumped her helmet occasionally as she went. She traveled this way, breathing through her mouth to avoid the smell, for many minutes, and after what must have been over a kilometer of slugging through disgusting sewage-ish sludge and pitch darkness, she felt the roof above her open up.

She was relieved to be able to stand fully upright again and she stopped to stretch and think for a moment. *How high up is the roof? Could there be a way out?*

She jumped, tentatively, reaching upwards, but didn't hit anything. She jumped a little harder the next time, and a bit harder the time after that, each time splashing back down into the nasty soup. On the try before she would have given up and soldiered on, something happened. As she was falling slowly back down, a light came on from somewhere above and to her left.

Splashing down again, she lost her balance and managed to get some of the mushroom-poop soup through the torn regulator hole and into her mouth. She regained her feet while spitting and gagging. Then she looked up towards the dim light. She couldn't see its source but could now see that she was in a trough as deep as where she had originally been dumped. She decided that she must have jumped high enough to activate some sort of motion-sensing light in the room above. This time when she jumped she also scrambled up the side of the trough, caught the rim, and hauled herself up. Sure enough, more lights winked on throughout the large empty room.

Dana worked the seals on her helmet, removed it, and shook it out, trying to dislodge the wet green clumps that had found their way in. She also did her best to wipe off as much of the green goop as she could from her spacesuit and gloves. She realized that the ham radio handset that had been attached to the side of her helmet was long gone — likely lost in the worm's belly. She did her best to suppress the combination of rage and panic that the loss of any way to contact her friends produced. Then she took a good look around.

It was another huge underground hewn rock room, similar in size to the one she had come from. This one, however, was almost entirely round, deformed only by the large opening onto the trough that Dana had pulled herself out of and, on the opposite side, a huge doorway into a large tunnel corridor that bent immediately to the right, so she could not see farther. The center of the room was completely empty, but the curving wall on her left was painted with some huge elaborate mural. To the right, the wall was lined with what looked like display pedestals in a museum. Each of the many identical pedestals had a different object on exhibit.

Part of her knew she should head for the exit as fast as possible and get on with the search for her friends. She started walking that way, but such signs of clear intelligence fascinated Dana and she found herself veering off to the right. Examining a pedestal at random, she was baffled to see that what it proudly displayed was, in fact, a golf ball.

Having a hard time believing her eyes, Dana picked up the golf ball and examined it. It had the familiar dimpled patterns, with a small picture of a flying crow on one side, and "Cro-Flite" was printed above "4" on the other. She returned it to its place and went over to the next pedestal, which supported two objects. Side by side were a large black and white bird feather and a quite normal-looking wooden-handled claw hammer. The next stand had a wallet-sized family photo showing a human man, woman, and two boys. Dana's mind reeled a bit as she recognized the artifacts left on the Moon by various *Apollo* missions. But what really shocked her was the next item on display: she immediately recognized the distinctive shape of the small lumpy indigo-colored rock. This had to be the same little meteorite that had pierced the Beetle on their trip from Earth, almost killing them. She knew she had left it hidden in Chip's water dish. If the bugs had this stone here, then what had they done with the Beetle and her friends?

From the notebook of Dana Dash

Dana reached out and picked up the small stone.

"AHEM!"

A voice broke the silence of the room. Dana whirled around. Still not completely used to the low gravity, her spin continued farther than intended and she had to stop herself and then turn back to the door. In the opening of the tunnel stood an adult woman in a tight green bodysuit.

At first Dana thought it must be Dr. Cavor, the only adult woman — or adult human being — that Dana had reason to believe was currently on the Moon. After a moment, the differences leapt out at her and Dana realized this was not only not Dr. Cavor, but not, in fact, a human woman. Her narrow eyes were a solid glossy black with no trace of white. What Dana had taken for tight clothing was actually something like snake scales. The woman had small antennae protruding from the top of her head and a thick, scale-covered tail trailing behind her, disappearing around the corner into the tunnel.

"Um... Hi," said Dana. "Uh... Do you speak English?"

"Yes," said the strange alien woman. "I speak many primitive and ancient tongues."

"Oh... good." Dana paused, at a loss for what to say or do. She felt like Goldilocks, caught in someone else's house uninvited. Although right now, she would almost rather it were three grizzly bears confronting her. The silence persisted, as the woman said nothing, staring with those unblinking obsidian eyes. Finally Dana said, "My name is Dana. It looks like I'm in your house, so this is awkward for me to be asking, but who are you?"

"I know who you are, girl on the moon. While I did not expect to find you here, it could only be you that I *would* find here. As for myself, I am the Grand Lunar, Queen of Selenites, Allmother. But you may call me Ms. Moon."

"Um... Ok... Pleased to meet you, Ms. Moon. This is a really nice museum you have here..." She wanted to say something to lighten the tension. Keep the alien talking while she figured out what to do next.

First she looked back to the weird collection of reverently displayed random objects. Her eyes fell on a pedestal proudly displaying a $2 bill, then the pedestal next to it with a clear plastic bag, tied off at the top. Nothing came to her mind to ask. *Is that a bag full of poop? I can't ask the queen that! Maybe I should make a run for it. Try to get past this alien-woman-thing.* She turned and looked at the large mural on the other wall.

"That's a really interesting picture," said Dana. "What is it?"

"That is chronogram depicting major events in the story of our people," said Ms. Moon.

"A chronogram?" asked Dana. "Is that like a timeline?"

"Time is not a line."

CHAPTER 23 — DOING TIME

Dana walked closer to the mural that Ms. Moon had called a "chronogram." The large picture, taking up a whole wall of the giant cavern, had a long, complex, flat oval border made up of one thick dark line with many thinner loops, sometimes crossing each other or nested inside each other, detaching from and re-attaching to each other. Each loop seemed to have a sidebar with text in a strange script Dana didn't recognize. On the inside of the oval border, lines extended, connecting to an inner ring of various pictures and more of the unreadable text. Inside that, Dana recognized the symbol from the tattoo on Dr. Cavor's hand and the Cavor School coat of arms. It was turned on its side, and at the point of each of the arrows that Dana thought of as the girl's limbs and her smile, there was a picture with more of the alien text with a line to part of the outer oval.

"I don't understand," said Dana. "How is this showing time?"

She looked at the outer oval border again. It could be like a timeline, if time were a great big loop. These attached pictures and text would be significant events. One of the larger pictures jumped out at her. It looked something like her vision in Dr. Cavor's office, which she barely remembered when conscious but still occasionally haunted her dreams. In this picture, the central figure, with a single bright green eye, reached out to touch two smaller figures who reached out for two more, even smaller, and so on, creating a fractal[83] effect. Surrounding this was a halo of orange like they were all on fire.

"It is probably for the best that you do not yet understand," said Ms. Moon. "Now, please don't make a fuss about this."

---

[83] *Fractals* are totally cool. If you don't know about them, look them up!

Puzzled by the last comment, Dana turned and saw that Ms. Moon now had two companions with with her. Two of the dust runner moon bugs, the type Dana had first seen on the Moon's surface, stood on either side of the queen.

"These are my personal guards, Phi and Psi. I am going to have them take you into custody. I know I can't kill you, or probably even hold you for very long, but there may be something I can do to change things. I will have to think on it. But first, I will have them search you, to make sure you haven't stolen any of my treasures."

Sorry, bug-lizard-lady, I found it first and I'm keeping it. Dana was still holding the small lumpy indigo stone in one hand, her helmet in the other. Perhaps because she had recently passed through the digestive tract of something larger than herself, a way to hide the meteorite immediately occurred to her. Turning back towards the mural to hide her actions, she quickly popped the stone into her mouth.

The two guards moved forward and her helmet was knocked aside to bounce across the floor as they seized her arms. Walking on each side, they brought her towards their queen, who bent down to look deep into Dana's eyes, then raised a hand towards the ceiling. Dana flinched, thinking Ms. Moon intended to strike her. Instead, one of the round lights detached from the ceiling and floated down to the queen's hand. Dana realized that the ceiling lights were also living creatures — buglike with wings and a brightly glowing tail section. Ms. Moon moved the giant lightning bug closer to Dana's face until she blinked and turned away from its brightness. The moon queen removed the bright light, then repeated the process several times, watching carefully.

"So interesting," said Ms. Moon. "Tiny sphincter muscles in your eyes adjust the size of the aperture based on the light level. Evolution produces some very strange results. Maybe I will try that one."

Then, without another word, Ms. Moon drifted backward, disappearing around the corner as the guards followed her with Dana in tow. Ms. Moon didn't turn, but backed around the corner, and Dana noticed that her tail retreated without bending. It was almost like the tail was moving the woman rather than the woman moving the tail. Dana puzzled on this as they followed the queen down the large tunnel, then gasped as they turned the corner to the right, and the reason for this strange mode of movement became clear.

The guards brought her into another large chamber. This one was filled with a giant long yellow-greenish sack, pulsating with an internal rhythm. At one end of it was a line of white spheres lying on the floor, each maybe two meters in diameter and, even as she watched, the tip of the giant sack opened up, and another giant sphere slid out onto the floor. *Eggs,* Dana thought. At the front of the sack was the rest of the body of the Queen of Selenites.

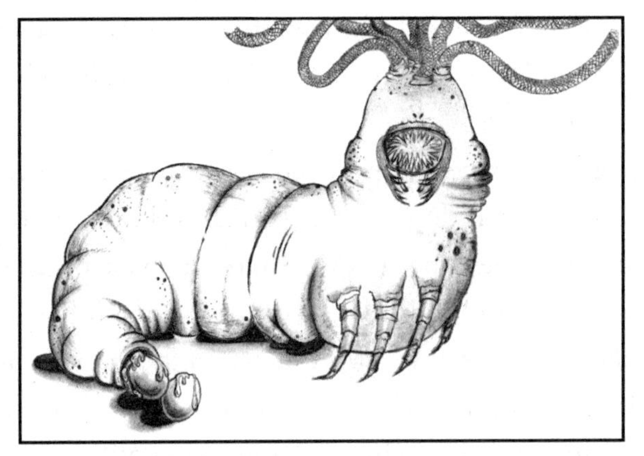

From the notebook of Dana Dash

Dana had seen pictures of termite queens, and this moon bug queen looked much like that. Her insect-like thorax and legs were gigantic, but they were dwarfed by the size of the abdominal egg factory. Given the moon's weak gravity, those giant bug legs might have been able to drag the rest of that massive body along one of the large hive corridors, but there was no evidence that they would ever need to. An army of smaller bugs serviced the queen, rolling eggs away and stuffing her gigantic mandible-surrounded mouth with globs of green jelly, carried to the queen in their own basket-like lower jaws. Her mouth was at the bottom of an eyeless head the size of a car. Where the eyes would have normally been sprouted were seven thick snakelike green scaled trunks, one of which Dana had mistaken for Ms. Moon's tail.

The Ms. Moon puppet, to which Dana had been speaking, rose up into the air on the end of one of these stalks, hanging above the great mouth like an anglerfish's bait light. The other stalks all disappeared off into other tunnels, singularly and in pairs, and Dana imagined other Ms. Moon figures at the ends of each those stalks, engaged in various activities. She wondered if she had been talking to the Ms. Moon who spent all her time hanging out in museums.

Suddenly, the alien strangeness was too much for her. Dana snapped. The stone, still in her mouth, prevented her from making the normal karate yell, but her body still did everything right. With martial arts skills drilled into muscle memory, she executed perfect grip-break rolls of each of her wrists simultaneously, reversing the situation so that she was the one controlling Phi and Psi's arms. Then she threw both of the bug guards forward. She might have only been ten, but she was a well-trained ten-year-old acting in about fifteen percent of the gravity her body was used to.

The two bugs tumbled through the air, straight towards Ms. Moon's massive munching mandibles. She didn't wait to see if the two guards were ingested, turning and bolting down the nearest tunnel that didn't contain any of the neck-puppet-people-body parts of the moon bug queen's anatomy.

She didn't manage to get very far. She started the long bouncing moon skip/run she had perfected on the surface, but before she had reached top speed, she heard a buzzing, whirring noise. Something was catching up with her fast. She turned and had time to make out that her pursuit consisted of two white flying discs spinning through the air. *Some sort of security drones?* Then one of them collided with her unprotected face and a jolt went through her whole body. Her muscles froze rigid and she crashed to the tunnel floor. The shock and the collision with the floor knocked the indigo stone into the back of her throat and she swallowed involuntarily. She just had time to worry about whether it was made of something poisonous, then think that the possibility should have occurred to her before she put it in her mouth, when the second drone also hit her. She blacked out.

When Dana came to, her head was pounding and she didn't open her eyes right away. Whatever she was lying on, it wasn't a material she recognized. It was spongy but firm with a bumpy texture. She was still wearing her space suit but the helmet and radio were gone. She had dropped it somewhere. Where? *In the moon bug queen's museum*, she remembered. No doubt it would go on display.

"Dana?" That was Noah's voice. She opened her eyes. Dr. Cavor and Noah were sitting on the orangish gray material beneath them and she was lying up against a wall. She assumed they were still underground — maybe it was the feel of something above her or the musty sour smell clinging to the air, or that there was air at all. When Bosco realized Dana was awake, he came to her side and nuzzled up against her arm, licking her chin happily. His tail whapped hard. She patted his head to signal she was okay before turning to her friends.

"What happened?"

"You should've seen it!" Noah said. His eyes gleamed with excitement. "After they took us down the hole, more bugs showed up. There must've been at least fifty! Dr. Cavor got the power restored and tried to take off, but there were too many of them holding on for us to lift by then.

"They were smart enough to wait until we passed through a sort of membrane airlock before they broke into the Beetle to get us. They must be at least somewhat intelligent!"

This was like pure gold to him, she realized. He had discovered a new phylum or even *kingdom* of life forms — new sentient beings — a whole new realm of possibility.

He continued. "They brought us through a tunnel system which they have to have dug out themselves. I think they live in a society similar to ants, but they're clearly much more complex and developed creatures. They dropped us off in this cell." He gestured around the room.

"They are definitely smart. I saw the queen. She spoke to me."

"What?" said Noah. "Really? None of them has said anything to us. The guards that tossed you in here looked like the ones that grabbed us from the Beetle. None of them talked."

Dana realized that it was only the queen who had spoken. "Right. Only the queen talked. The guards were silent."

"She spoke English?"

"Yes. She said that she spoke a lot of languages." Dana looked around. "Oh, and she had seven heads."

"What?" Noah gasped. "Really?"

"Well, kind of heads. The necks were really long... and the one I saw had a body on it, so maybe not really necks..."

"What?" said Noah. "Back up. What did she look like?"

"A termite queen with hydra heads. Remind me to draw her for you later." Dana looked around at the small room they were in. The circular room wasn't very far across, but the ceiling was high. The dark pinkish walls were covered with a regular dimple pattern but looked too smooth to climb.

"So, Dr. Cavor, I guess we're in Space Bug Jail?" she asked.

Dr. Cavor was seemingly deep in thought. She stroked her chin like she was playing with an imaginary beard.

"Do you see that hole in the ceiling?" she said to Dana. Dana nodded. There was a small opening in the ceiling, maybe big enough to put her head in. "We were dropped in through that opening, and then it got smaller. I believe I have an idea how to get it to open up again. I think we can all jump high enough to grab the rim in this gravity."

"But what about Bosco?" Dana inquired.

"And the security drones?" Noah added. "They're posted all around the room up there," he explained to Dana.

She was instantly annoyed that Noah had seen the layout above and she had not. She wished he had been knocked out instead of her and then immediately regretted thinking it. Noah was her best friend. She couldn't wish him to be knocked out like it was no big deal, plus he had been so nice this whole trip. He was always nice — to her and everyone else. Why was she so insanely competitive with him when he was supposed to be her friend and work partner?

Dr. Cavor said, "It will be risky, but we do not have any other way out. If you do not feel up to it, I completely understand. I would never make you do — "

"I'll do it," Dana volunteered. "Give me a few minutes. My head is still spinning and I feel a bit queasy. Talk about something. The sound of your voice helps me keep my mind off my head and stomach."

"What should we talk about?" asked Dr. Cavor.

"Tell us more about dimensions. Three space dimensions, one time, and how many others for the various forces?" asked Dana.

"Hey," said Noah, "is time really a fourth dimension?"

"Well," said Dr. Cavor, "the best answer I can give to that is a definite maybe. All we have are models. And it is important to remember that models help us explain and predict reality, but they are not real. All models are wrong but some are useful.[84]

"Let us consider dimensions to see how different ways of looking at things can be equally wrong, but also useful in different ways." She pulled a notebook and pen from somewhere in her clothing and began sketching.

"In the pure mathematical sense, a dimension is a number that describes something — anything, really. Here, we are describing position with X, Y, Z coordinates."

[84] This quote is generally attributed to statistician George Box (1919 -2013). But it is also worth remembering that some models are so far less than useful at predicting reality as to be suicidal. So do always try to make sure that all of your wrong models are as right as you can possibly make them.

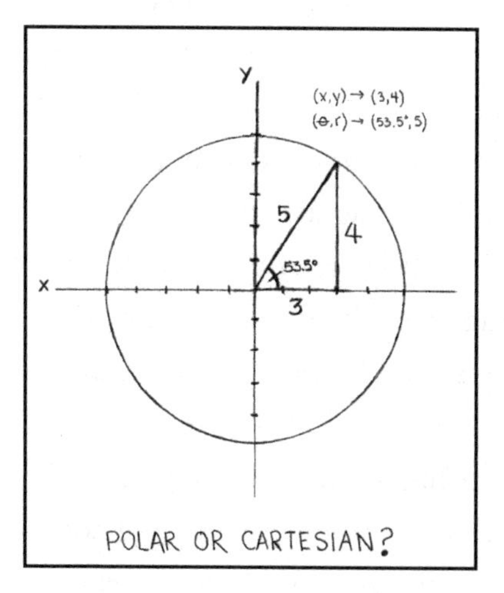

From the notebook of Catherine Cavor

"Right," said Noah. "Dana was telling me that this is how the Dash family does Easter Egg hunts."

"Very well," said Dr. Cavor, "so you know Cartesian[85] coordinates. If you want more dimensions... certainly, you could have time. But you could also measure temperature at any point. That could also be a 'fourth dimension.'

"Try to remember that a new dimension is simply another number that describes something that the previous dimensions could not. It does not have to be related to the other dimensions. Even if two numbers are helping you measure the same thing, such as position, they do not have to be the same type of numbers. Another common way of measuring position is to use angles and distance from the center."

"Oh!" said Dana, "Are we going to do polar coordinates?"

"Exactly!" said Dr. Cavor, starting a new drawing. "It still takes two numbers to map a point in two dimensions, but we can do it with numbers that have very different meanings. The X, Y point 3,4 can also be located

---

[85] This X, Y, Z coordinate system is named after René Descartes (1596–1650), the philosopher, mathematician, and scientist who invented it.

by degree and radius length as approximately 53°,5."

"Ok," said Noah, "I understand how polar coordinates work, but time can't be polar? Can it?"

"Ah!" said Dr. Cavor. "Not only can it, but that is the way you normally think of it. Here, let me show you Cartesian time..." and she drew two circles which she labeled *t1* and *t2*. "This is your polar time at two moments, expanding from the Big Bang." She drew what would be the x and y axes but she wrote *Tx* and *Ty*. "And these are two cartesian time axis. Flattened from four space, of course.

"As for models being useful, Cartesian time can demonstrate special relativity effects. This tangent line is your light cone, obviously." She continued to add details to the drawing almost faster than the eye could follow and her speech actually seemed to speed up, even as what she was saying made less sense. "So you see, moving from point A to point B is actually moving from A*1 t1* to point B*2 t2*, which stays within the light cone. But A*1 t1* to C*2 t2* cuts back in time through moment *t1*..."

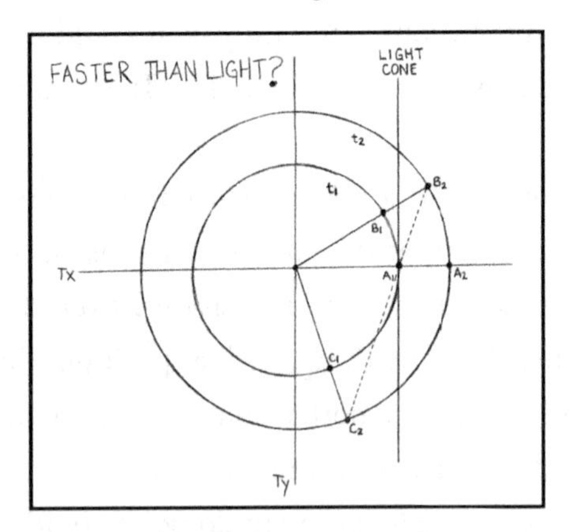

From the notebook of Catherine Cavor

She looked up. Dana and Noah exchanged bewildered glances, and then looked back to her. Only Bosco had his head cocked to one side as if still listening attentively. Then he kicked himself in the head to show he had only been preparing to scratch an itch.

"Ah yes," she said. "Geniuses, certainly, but only ten-year-old geniuses. Perhaps we should save this material until at least next year... Let us just say that four Cartesian time values can locate any point in space-time as accurately as three space and one polar time value can."

"Um, ok... we'll say that." said Noah.

"Wow!" said Dana, "So time really isn't a line... Now my head is spinning for an entirely different reason, but I think I'm feeling better physically."

"Very good," said Dr. Cavor. "Then what do you say we take a break from theoretical physics, as stimulating as it is, and free ourselves from captivity?"

CHAPTER 24 — JAILBREAK

Dr. Cavor said, "First things first. Once we get out of this cell, we may need to communicate if we get split up."

"I lost my helmet when I got caught," said Dana. "And one of the ham radios is gone too."

Noah spoke up. "The other ham handset is still in the Beetle anyway. I think we will have to stick together."

"Do you both have your mobile phones?" asked Dr. Cavor.

Both children indicated they did. "But how can that help us?" asked Dana. "Our phones can't work on the moon. They need to connect to a phone company cell tower."

"Not a problem," said Dr. Cavor. "We need to make them work with a different wireless carrier. Welcome to CavorTel — great lunar coverage and very reasonable rates, certainly better than the roaming charges your old phone company would make you pay up here..."

She gathered the children's phones and took out her own, playing around with it for a few minutes. She explained as she worked.

"I am setting up my phone to act as a base station. Our phones will work as normal. Of course, we will only be able to dial each other."

A minute later she proved it by dialing Noah's phone from Dana's phone. Then she called herself and conferenced the calls together to test it. "There. Now we can reach each other in case we're separated." She ended the connections.

"You really have to show me what sort of apps you have on that phone of yours some time," said Dana.

"Perhaps when we get back." Dr. Cavor handed back both of the children's phones. "Now, let us see if I can get us out of here. That opening up there gets larger to let people in and smaller to trap them.

Perhaps I can open it up from here. With a little help from Mr. Bosco — if you would?"

Bosco padded over to Dr. Cavor, who proceeded to rub his fur vigorously. "We are fortunate that you are here, sir. Give me all those extra electrons. Excellent! Dana, you first, get ready to jump. Ready?" When Dana nodded, Dr. Cavor took her hands off the happy dog and reached down to touch the floor of their cell.

Dana heard a tiny pop and saw the spark in the dim light. It seemed like such a small effect to cause what happened next. The whole cell shook. The floor they were standing on heaved and fell, and smooth white spikes popped out of the indentations in the walls, which rippled like they were alive.

Alive! Dana realized that, like seemingly everything else in this strange place, their strange jail cell was a living organism. With teeth! Dr. Cavor had done something to deliberately annoy a giant creature while they were trapped inside its mouth. Dana wondered briefly how many times today she was going to be eaten alive.

"Now, Dana! Jump!" Dr. Cavor was looking upward. Dana followed her gaze and saw that the opening at the top of the jail cell mouth tube was convulsing too. Open and closed, but mostly open.

Dana didn't hesitate. She leapt upward, what would be an impossible distance on Earth, and caught the upper edge of the floor above. As she scrambled out she could see where the round hole in the stone ended and the giant thick rubbery lips of the jail cell monster began. She turned around and quickly extended her hand back down into that horrible mouth. "Noah! Jump!"

Noah leapt upward, his hand outstretched, and Dana grabbed his arm, pulling him to continue his jump into one smooth motion that left him standing on the stone floor next to her. Before Dana could even turn back to the hole, Dr. Cavor shot up out of it, Bosco held in her arms. Dana had picked her dog up before and it was not, by any means, an easy feat. How Dr. Cavor had done it so easily while jumping, even in low gravity, was beyond her.

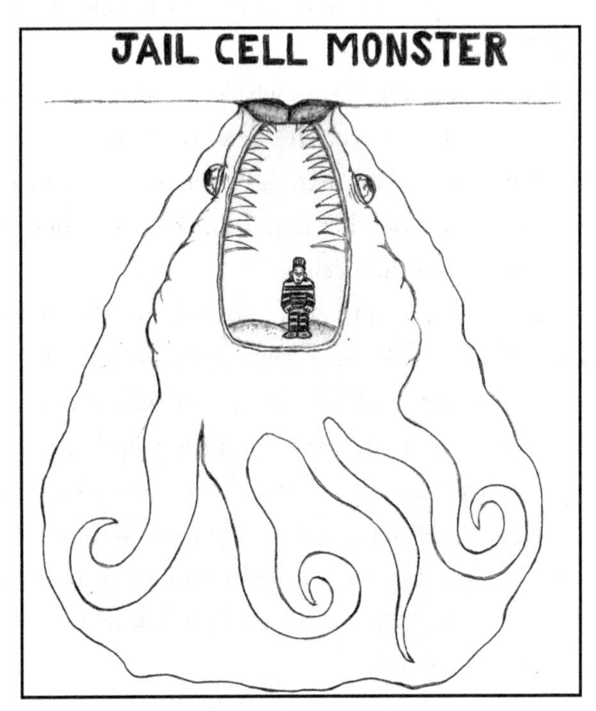

From the notebook of Noah Knight

While Noah was looking back down the mouth-hole they had emerged from, a look of horror on his face, Dana finally had time to look around her. They hadn't been kidding about security drones. Several of the white, disc-shaped objects, the things that had chased her and knocked her out, were levitating at various points around the room. Dana's fingers itched to take one apart and get a firsthand look at whatever this alien technology was like, but she restrained herself. She knew they were dangerous.

"Come along, children. These drones do not look happy with us." The spinning discs had come to life and were all heading towards them. Dr. Cavor dodged and pulled Dana along with her. Noah dove in the opposite direction. Bosco yapped at the drones like they were some sort of game. Dana was reminded of playing fetch with him and was struck with a sudden idea. The drones actually did look quite a bit like his favorite Frisbee...

"Bosco! Go fetch!" The Labrador happily complied. He sprung off his hind legs and tackled a drone out of the air. His powerful jaws snapped the disc in two. "Good boy!" Dana said, then "Ewwwww!" when she saw the insides of the disc exposed.

It was not a machine, as she had assumed. It was, like the lights, the jail cell, and everything else here, a living organism, full of guts and goo, which Bosco dropped at her feet, tail wagging madly. But she managed to temporarily put aside both her disgust and her sympathy for all living things to point and shout "Fetch!" again and again.

If the drones were shocking him the way they had her, the dog seemed somehow immune. Soon Bosco had successfully destroyed all of the security drones and left the remains in an oozing heap.

Noah smiled in relief and said, "We're lucky to have you here, Bosco!"

"Yeah," Dana agreed. The dog cheerfully rubbed up against her as if saying *Yeah you are! Are you only now realizing that?*

Dr. Cavor smiled tightly. "Well done, all of you, but we have to get out of here now, before more of the larger security bugs come."

"I'm pretty sure that's the way we came in," said Noah, pointing at an exit tunnel. He started walking that way.

"Wait!" called Dana. "Look over here!" She had spotted a number of items on the floor in a little nook off of the main jail cavern. It was a collection of random items that the bugs had taken from the Beetle, perhaps as gifts for their queen's museum: Dr. Cavor's amazing gravity-reversing machine, Chip's cage, and the crowbar.

Dana grabbed the crowbar and also picked up Chip's cage. The little fox looked ok, if a bit disheveled. Thankfully the bugs had left him alone inside the aquarium. Dr. Cavor grabbed her machine and tucked it under her arm.

"Ah, yes, I might need this." Then they both followed Noah's lead, scrambling to get as far away from the jail as possible.

They hurried down the cramped tunnel, but when they came to the first intersection, no one knew which way to go. "I guess we have to pick one at random," said Dana. She started forward.

"Wait," Noah said, stopping her. "I have an idea."

They had been walking for about ten minutes now, and Dana was feeling exhausted from the day's ordeals. She was happy to stop.

Noah continued, "These tunnels go on for miles, but I bet I know how we can find a route that goes directly back to the Beetle! These things really are similar to ants! See this?" He pointed to a line of shimmery clear liquid along the tunnel floor.

Dana gasped and continued Noah's thought, "Ants find their way home to the main nest by marking their trail with a natural chemical. These space bugs do the same thing! It's a chemical scent trail!"

"Yes," Noah agreed. "Do you smell that musty thing?"

"Yeah." Dana had thought it was the contained oxygen. She was annoyed that Noah had beat her to this idea. "There were *a lot* of space bugs carrying us to the cell," she realized. "They probably left tons of this stuff! We have to follow the freshest trail, for lack of better words."

Noah beamed. "Exactly!" The two kids continued forwards, followed by Bosco and Dr. Cavor. At each choice of which tunnel to take, Noah checked the density of each chemical trail they found. Some were hardened, but a few were still moist, as if the space bugs had recently passed by — which was not at all a pleasant thought.

They came to a tunnel where the floor had been completely coated with fresh chemical. Noah's face lit up.

Dana knew exactly what he was thinking, "We are not taking samples of this gunk." She herself wanted to stuff as much of the moon into the Beetle as could possibly fit. But she knew that was unrealistic. They could always come back to the Moon. That thought made her heart pound in her chest. Would Dr. Cavor let them return after all of this? "Come on."

She grabbed Noah's sleeve and they followed the trail up through the tunnel. The floor they walked on was wet with pheromone produced by the moon bugs. She was thankful she was still wearing the space suit pants that covered her shoes. Noah and Dr. Cavor were getting their shoes soaked in bug juice.

Noah nudged her as they walked and said, "How's your head? Did they hit you? I couldn't tell."

"I don't think I have a concussion, if that's what you're asking. But it does still ache a bit. If they hit me, it was after I was already unconscious. It was an electric shock from one of those drones that knocked me out. Like an electric eel."

Bosco had been running up ahead and then back, but as they neared what looked like an opening up of the tunnel into a space ahead, he hung back behind them and whined.

"He senses something," said Dr. Cavor. "Move forward cautiously."

As they slowly approached the end of the current tunnel, small clicking sounds they'd previously assumed to be from other passages grew louder. Their tunnel opened out into a huge cavern filled with space bugs. Dana mentally kicked herself — she should've noticed the dried pheromone mixed in with the new. It was the most heavily used path because it led straight to the hive, not because a bunch of bugs had recently traveled down it with Dr. Cavor and Noah in custody.

The opening of the tunnel they were coming in from was maybe five meters above the floor of the cavern. Moon bugs swarmed on the floor of the large space and were camped out in small sleeping caves lining the walls. Separating one row of caves from the other were bridges winding up the walls in a spiral formation. The bugs bustled about, transporting food and supplies. Noah gasped and Dana could see his internal struggle between his scientific curiosity to stay and observe the bugs and his sense of self-preservation, which was probably telling him to run for his life.

Suddenly, Noah stage-whispered, "There it is!"

Dana turned to look where Noah was pointing and saw their spaceship. The Beetle was up on a wide ledge, maybe a quarter of the way around the cavern from them and at about the same level as they were, where a couple of other tunnels entered the huge hive. Looking down at the cavern floor,

Dana considered the sea of moving bugs between them and their only hope to ever see home again.

"It's impossible," she hissed to her companions. "There's no way we can get across to the Beetle without them overpowering us first."

"Hmmm." Dr. Cavor considered the scene. "I think what we need here is a distraction." She looked back and forth. "Children, I am going to head for that other tunnel, making a ruckus." She pointed in the opposite direction from the Beetle to a much closer, very large exit. "If we are fortunate, I can get them to follow me. Once they do, all of you will sprint to the Beetle and see if you can fly it up the larger of those two tunnels. Keep going up wherever you can, and try to find the way they brought us in."

"What about you?" Dana interrupted. "We aren't leaving you behind. There's no way you will be able to get back to us."

"Trust me. It will all work out the way it should. It always does." Before Dana could object further, Dr. Cavor had launched herself off the side of the ledge they were on and was running expertly down the side of the wall.

Bosco started to follow her, but Noah grabbed his collar, saying, "Whoa boy, you're staying with us."

Dana and Noah watched in amazement as Dr. Cavor reached the first surprised moon bugs and kept going right through them, knocking them aside, running across their backs, and vaulting over them.

The assaulted bugs cried out in a weird warbling tone that was taken up by those around them until the whole cavern thundered with it. Dana could see the surface bugs, jailer bugs, and other types she had not seen before all start heading in the direction of the fleeing school administrator. It was working.

"It think she really is going to make it to that tunnel," said Noah. Dr. Cavor made a leap that could not be humanly possible, even in the weak lunar gravity. She flew over the remaining moon bugs between herself and the large tunnel, disappearing into its mouth on the down arc.

Dana looked the other way, towards the Beetle. It was a lot clearer path than it had been. The bugs there had all dropped what they were doing and moved to pursue, or at least watch what was going on. She had no idea how long to wait before making their move, but then she heard something in the tunnel behind them and turned to see a group of jailer bugs coming up their tunnel, no doubt in response to the commotion in the hive.

"Time to go!" She gave Noah and Bosco a push in the right direction and then followed them off the ledge into the fray below.

CHAPTER 25 — BREAKING THE GLASS CEILING

Dr. Cavor's distraction had worked perfectly. Dana, Noah, and Bosco ran at their best skip-jumping speed across the floor of the hive cavern, and none of the moon bugs seemed to notice them until they were halfway across. By the time they reached the sharp slope up to where the Beetle was parked, pursuit was almost right on their heels, but they were going to make it.

Then, as they bounced over the crest, a few meters from their little spacecraft, one of the white taser discs was there, right in front of them. Dana cried out as it zoomed towards her face. In a blur of brown fur, Bosco was in front of her and the white disc was crunching between his teeth. She saw it spark and flash as it died.

"Good boy!" yelled Noah. "Thank goodness he seems immune to their electric shock."

"Yeah," Dana yelled back as she dove between the Beetle's legs and opened the hatch. "My dad tried an invisible fence to keep him in the yard when he was younger but it didn't work on him at all. He basically wanders at will but he always comes back."

Dana dropped the crowbar and climbed up to the controls to power up the Beetle. Bosco jumped in and Noah set down Chip's cage briefly to close the door, then returned him to his normal place in the spacecraft. It didn't look like the bugs had done too much damage inside the Beetle. Dana quickly activated the correct shutter and the little spacecraft lifted — not a moment too soon, as the moon bugs were arriving. One leaped, and managed to grab hold of the Beetle's landing plate as they rose. Then another jumped to grab hold of the first. They were making a living chain, trying to drag the Beetle down.

Bosco was squeal-barking like mad at the clinging bug and it looked like they were not going to make it. Dana had the airlock ceiling shutters fully open, exposing as much of the Beetle-heart to the moon as possible, but more bugs were climbing the chain and the combined weight was too much to lift. They would be dragged down and overwhelmed in moments. What could they do?

Dana abandoned the controls, leaving them on full lift, and was past Noah and into the airlock in moments. Picking up the crowbar, she opened the outer airlock door and slid out onto the landing plate.

The first bug coming up over his friend's back got a smack between its three eyes with the curved part of the bar and fell backwards. Then she was using the sharp flat end to pry under the bug's hand where it gripped the landing plate. It had too many hands, and the bug would easily grab back on whenever Dana shifted her attention to the next hand. Another bug came over the top. She hit it, and the cycle repeated..

Finally, the weight of all the bugs, the pull of the Beetle's neg-grav core, and Dana's ongoing efforts with the crowbar were too much for the attackers. The clinging bug at the top of the chain lost its grip, and the freed Beetle, no longer weighed down, shot skyward.

"Wow. That was close," said Noah as Dana climbed quickly back past him to regain control of the ship.

"Not out of the woods yet," said Dana as she got back to the controls. The Beetle was rising fast and still accelerating. "Not by a long shot." She shuttered the neg-grav and the Beetle started to slow, but not quickly enough. The roof of the cavern grew nearer and nearer. One particularly large stalactite was pointing down right at her. Dana watched through the clear roof above her, frozen. *This is it. The Beetle is going to end up pinned like a butterfly.* But they came to the top of their climb just short of being skewered by the stone spike. The tip of the stalactite tapped the clear polycarbonate lightly, between two of the top shutter slats, and then the Beetle started to fall again.

Dana regained control of their ship and steered it down and towards the center of the cavern.

"Where are you going?" asked Noah. "Dr. Cavor said to go up that wide tunnel."

"There's no way we're leaving her behind," said Dana. "If she doesn't like it, she can yell at me all she wants after we save her."

She modified the shutter settings and they headed towards the tunnel Dr. Cavor had taken.

Since Dr. Cavor had done most of the flying until now, Dana's control was far from perfect, but she was getting the hang of it pretty quickly. The Beetle zoomed down and across the cavern, back the way they had run and then farther, picking up speed the whole way. Dana didn't even slow the craft as they plunged into the tunnel that Dr. Cavor had taken.

This tunnel wasn't nearly as large as the one they were supposed to have taken. It was a tight fit to begin with, but the bottom part of the tunnel was full of moon bugs — they were still piling in from the large cavern to join the pursuit. Staying below ceiling level, Dana zipped along above the heads of the bugs. At one point, the top of the Beetle bumped into the tunnel roof and the craft bounced downward onto the backs of moving moon bugs, but their forward momentum carried them on before any of the bugs could try to climb aboard.

The tunnel was angled upward, so at least they were going in the right direction. Finally they zoomed out into the bottom of a dimly lit crater. It was almost lunar night. A tiny slice of the upper part of the crater rim was reflecting enough light to make their surroundings barely visible. Dana brought the craft a little higher to give them some distance from the bugs below them where they still streamed out of the tunnel.

Noah said, "I think we're in some sort of alien farm."

He was right. They had come out into a giant crater clearly used for agriculture. Rows of strange vegetables and spindly trees bearing plump spiny orange fruits surrounded them. There was no obvious irrigation system, but the ground looked wet.

"I saw something like this earlier," said Dana. "But it was outside, with no atmosphere."

Was there atmosphere here? They hadn't passed through any sort of airlock. But what would be holding the air in? She glanced up at the stars above her. A tower of stone rose up from the center of the crater they were in. Dana followed the tower's path with her eyes. Oddly, it got wider as it went up. At the top, she could make out some sort of transparent roof spanning the whole crater. The central stone tower seemed to have a flat top that almost matched the height of the crater walls, and the see-through roof was supported by it. The tower seemed to be made of the same melted stone as the crater walls, and Dana wondered briefly how a meteor strike could have formed this large deep crater but left a standing tower of stone directly in the middle.

But more importantly than any of these speculations, Dana realized that she was seeing Dr. Cavor. The teacher was climbing the stone column that looked like the Space Needle and was almost at the top. A stream of pursuing moon bugs was clawing its way up the column right behind her. They would reach her in minutes, if not seconds.

Dana pointed, "Look! There she is." She steered the Beetle up towards the top of the stone tower and they drew up beside Dr. Cavor as she was nearing the top. At first she wondered how this incredible woman was climbing while still holding the gravity-reversing machine. Then she did a double take. "Am I seeing what I think I'm seeing?"

Dr. Cavor was able to use both hands because she had stuck her left arm all the way through the strange machine and was wearing it. This left her with two hands to grip the rock column, but there was something weird about those hands. As she reached the top of the moon-needle, Dana realized what it was. She had two right hands. Identical. Thumbs on the same side. The machine had reversed her arm! Yet she was climbing with it like that was of absolutely no matter.

Dr. Cavor looked over at them as they hovered close by in the Beetle. Dana made motions that she should try to climb onto the base plate but

Dr. Cavor shook her head. Somehow holding onto the melty-looking rock with one hand, she pulled out her phone.

Noah's phone rang and he pulled it out of his pocket.

"It's Dr. Cavor!" he said excitedly, as if it could have been anyone else calling. He answered and put it on speaker. It was a video call, which hardly seemed necessary, as they could see each other clearly through the open shutters.

"Dana, Noah! Thank Betelgeuse[86] you are safe!" said Dr. Cavor. "But I told you to go the other way. There is no airlock here. You have to go back."

"We can worry about that once you're in here with us. I'm coming to get you," Dana insisted, and drifted the little spacecraft even closer.

"No. I have something to do. And you cannot be here when I do it." She had reached the barrier above, and rapped on it with the phone. "But I do need to get higher. I guess you probably can and should go out this way now. Do as I say. Do not panic. No matter what happens."

"What are you talking about? Get in the Beetle! Those bugs are almost up to you."

"No. You keep going. Let me just get this thing out of your way."

"What do you mean? What are you doing?"

Dr. Cavor tucked the phone in her pocket. Holding onto the rock with one of her right hands, she held her other right hand out. The hand glowed blue for a moment before the woman struck with blinding speed, her palm hitting the central stone pillar sharply. Dana heard a sound like a giant gong being struck and thought she could see a shockwave travel outward through the glass. Bosco howled back at the noise as the wave of force reached the edges of the crater and cracks formed there, travelling back inward from the edges.

---

[86] *Betelgeuse* is a star, a red supergiant, in the Milky Way Galaxy. It's one of the brightest stars in the night sky, and looks distinctly red. It's a lonely star, and unlike our sun, it doesn't have any planetary friends in orbit. It's pronounced "beetle juice," and you can say it as many times as you like.

"What is she doing?" cried Noah. "There's no air on the other side of that."

Then the glass ceiling shattered above them.

Dana expected the glass to shower on them in giant shards. But this was based on living all her life in a place where gravity was higher, and explosive decompression was not a normal risk. Instead of crashing down, the glass blew outward with a surge of hurricane-force wind. The air inside the dome suddenly had someplace new and interesting to explore and decided to take the glass with it.

Everyone in the Beetle was thrown to the ground as the little spaceship was sucked upward but also violently sideways, underneath a fountain of spinning glass shards that glittered in the lunar sunset, rising higher at the edges like an umbrella turned inside out by a stormy gust of wind. Things inside the little ship pinged and popped and creaked ominously, as the aluminum framework was stressed by suddenly having to, once again, hold its pressure against an outside vacuum. But the little spacecraft held together.

Bosco whined as Dana raised herself up from the deck to the Beetle brain control panel.

Noah looked out the windows, scanning in all directions, shielding his eyes against the last sliver of sun setting in the west. "Where is Dr. Cavor?"

Dana looked around too. Glass, dirt, ripped-up fruit trees, and flailing moon bugs hung in the sky around them in all directions. But there was no sign of Dr. Cavor.

Then they heard her speak. "I am right here."

Dana and Noah looked around wildly. How could she have gotten into the Beetle?

"Look down," said Dr. Cavor. "On the floor. I am still on a call with you."

Noah retrieved the phone and held it up so both of them could see. Dr. Cavor was holding the phone at arm's length so they could see she was standing on top of a roughly circular piece of the glass, still attached to the top of the moon-needle-stone-column. Her hair had come loose and she had no longer wearing her glasses. Dana could see her eyes clearly now — fiery, shining unnaturally bright, like her eyes were portals to another dimension themselves. The glass beneath her feet extended a few meters out from the stone, which was temporarily stymying the moon bugs. But Dana could see they were in the process of building living scaffolding to get around the barrier.

Dana wondered how Dr. Cavor had got herself out around that edge. Then something else occurred to her. "How are you breathing? How are you talking? All the air is gone."

"I am not breathing. And I am conducting my voice into the phone through vibrations in my hand," said Dr. Cavor, as if this was a perfectly normal and obvious explanation. "But that is not important right now. You need to — "

"What are you?" asked Noah. "Are you one of the aliens? Xenos?"

"I don't care what you are," said Dana. "We're coming back for you right now."

"No! No. Do not worry about me. This is all part of the plan. I promise you will understand everything eventually, but right now you need to do exactly what I say. Take the Beetle back down near the surface. Get outside the rim of this crater. Stay low but get as far away as you can. The edges should be high enough that you will be safe."

Dana took stock of their position. The air escaping from the roof had blown them sideways and upward. They were almost outside the rim of the crater now and they hadn't been thrown all that high.

"Safe from what?" asked Noah.

Dr. Cavor brought the phone closer and pointed it at the machine she was still wearing on her arm. On her wrist was her watch-like bracelet, with the bubble-covered red button instead of a watch face. The top of the button was lit with red glowing figures that were changing as they watched. Noah looked confused for a moment, then turned his phone to get the proper angle. The numbers were a countdown timer and it had just passed thirty seconds.

"Children, remember I told you what happens when antimatter meets regular matter?" She turned the camera back to her face. "You need to get on the other side of that crater rim right now!"

Something about Dr. Cavor's voice told Dana that this was for real. She put the Beetle on a course towards the crater rim.

"You're blowing yourself up?!" Noah was clearly upset about this. His left hand tugged madly at his shirt. "Why on earth would you do that?"

Dr. Cavor laughed. "I would not, on Earth," she said. "Again, you will understand later. But I will say, this is an opportunity to take out some of the Xenos before they can hit us, now, while we still have a few more years of protection."

"And it's worth killing yourself?!?" yelled Noah.

"Years of protection?" asked Dana.

"No more time, children."

On Noah's phone screen they saw that the moonbugs had made it out and around the edge of the glass pedestal and were coming at Dr. Cavor from behind. Dana was about to yell out a warning when Dr. Cavor crouched down, pulling the phone in close to her belly as she curled into a ball around the machine she still wore on her reversed arm. All that displayed on Noah's phone screen now was the glowing timer on the red button. Then they saw Dr. Cavor's index finger flip up the plastic guard over the button.

"Good luck, Dana," said Dr. Cavor. "Although you certainly do not need it. You are the girl on the moon now — or close enough not to matter."

The timer was finishing its countdown. 3... 2... 1... They saw Dr. Cavor's finger press the button...

...and absolutely nothing happened.

She pressed it twice more in rapid succession before the phone call suddenly dropped, leaving the kids wondering what had gone wrong. Dana looked away from the screen and over her shoulder back towards the crater.

They were now below the level of the crater rim and out of line of sight with Dr. Cavor, which explained the dropped call, but not why the supposed antimatter bomb hadn't worked.

Then the lunar dusk exploded into light. Dana's eyes were saved from the direct light of the blast by the crater wall. But the white-hot light, from the geyser of glowing debris spewing into the sky, forced her to shield her eyes. Then the protective crater wall was obliterated and a wall of white-hot solid, liquid, and gaseous moonrock cascaded toward them. It didn't matter what Dr. Cavor really was; nothing could have survived that blast. Dr. Catherine Cavor was dead.

And when that blast wave reached them, they would be too.

CHAPTER 26 — PHONE A FRIEND

Dana leapt to put the Beetle in motion, zipping away across the lunar surface at maximum acceleration. With no air resistance to fight the build-up of speed, the Beetle had no real upper speed limit. If the wall of debris didn't reach them in the next few seconds, before they were moving faster than it was, they would be all right.

There was no mushroom cloud, just an outward blast of ejected material. Without atmosphere, the dust didn't hang in the air, but the moon's lower gravity meant that the debris would be flying a lot farther. They were outside an expanding ring of glowing molten rock, tiny particles, and gas that moved up and out. The ring was higher above them, as most of the blast had been directed upward and outward by the shape of the crater. The raised crater rim had temporarily held the blast wave back at ground level — enough that, skimming along the surface, they had not yet been hit by the wave front.

The wall of churning debris got closer and closer, until it would have almost been close enough for a rider on the outside of the Beetle to reach out and touch. "It's going to get us!" yelled Noah. The panic made his voice squeaky and Bosco whined loudly.

Then it stopped getting closer and Dana let out a breath she didn't know she had been holding as they began to outpace it. They had successfully outrun the blast. "I think we're ok. Or at least we should be. Some of this stuff probably has escape velocity,[87] or near to it, and won't come down again. I'm going to maintain a slow rise while moving along the ground."

---

[87] *Escape velocity* is the speed at which an object will never return to the ground because it is going too fast for gravity to stop it before it gets too far away. This is possible because the force of gravity decreases with distance.

“Ok, whatever you think.” The panic was not yet totally gone from his voice, and he suddenly pointed at something and said, in alarm, “What’s that?”

Dana looked, and saw another plume of dust and debris off to their left, then a third, a little past that. “The blast wave must have gone down the tunnel into the hive. Those must be other tunnel exits.”

“Do you think she killed them all?”

“No idea,” said Dana. “But if not, it certainly wasn’t for lack of trying.”

While Dana continued to pilot, Noah got them each some water and gave some to Bosco and Chip. There had been breathable air in the bug tunnels, but the humidity must have been extremely low and they were quite thirsty. They had tracked lunar dust, bug pheromones, and green cheese fungus into the Beetle and the combined smell was so strong Dana could taste it in the back of her parched throat. She drank a normal full day’s worth of water in a series of long chugs before feeling better and saw Noah do the same. Then she peeled off her space suit and had Noah fetch a large plastic bag to store it in, which hopefully would reduce the fungus odor.

After traveling an unknown distance, Dana decided they were far enough away. She closed the top front shutters and opened up the rear ones, applying the brakes until the lunar surface finally stopped streaking by below them. They were hanging above the lunar surface as close to stationary as she could hold them. Then she closed all the shutters and tried to calm herself.

“Well, we’re still alive,” said Noah, “That’s the good news.”

Dana replied, “And the bad news is, we’re still on the Moon, Dr. Cavor is dead, and if any alien-moon-bug-monsters are still alive, they’re almost guaranteed to be extremely unhappy with us.”

“Right, that’s the bad news as I see it too. So let’s think about what good news we can make. What can we can do to improve our situation?” Noah had calmed down and was back to his usual level-headed self.

"Well, the only thing on that list that we can do anything about is that we're still on the Moon. So let's figure out how to get home. I assume that it's possible — Dr. Cavor seemed to think we'd be able to get back."

"Why not?" said Noah. "Open a shutter to the pos-grav that points at Earth, open a shutter to the neg-grav that points at the Moon, and we're on our way? No?"

"Yeah, but it's not going to be that easy. Remember what Dr. Cavor said about side velocity when we left Earth? We had a velocity from the spin of Earth that would be enough to put us in an orbit."

"Right, but the Moon is smaller and rotates more slowly. That rotational effect can't be much here."

"Yes," said Dana, "But that isn't the problem." She was turning away from Noah in the near-freefall environment of their entirely gravity-shielded habitat. She put out a hand to grab a strut and turn herself back to face him. "The problem is that we're now on the Moon, which is orbiting the Earth. So we have all the speed of that orbit. When we were intercepting the Moon and decelerating into it, we were actually building up orbital speed by pushing against the Moon. Slowing down relative to the Moon's speed, so we could land, meant speeding up perpendicular to the Earth. We were matching the Moon's orbital velocity around the Earth.

"Here, let me show you on paper," she said, pulling a pen and her notebook out of the utility cabinet at the top of the craft. She drew two circles, one representing the Earth and then a bigger circle through the Moon and around the Earth representing the Moon's orbit. Then, with pen to paper, she said, "Here we're on the surface of the moon. Our velocity matches that of the Moon going around the Earth." She drew an arrow in that direction. "Now if we push off the Moon directly towards Earth and we do nothing to cancel this side velocity, then we will miss the Earth. Instead of getting home we will put ourselves into some new Earth orbit. And if we miss I don't know how long it will take us to get back. Or if we ever get back!"

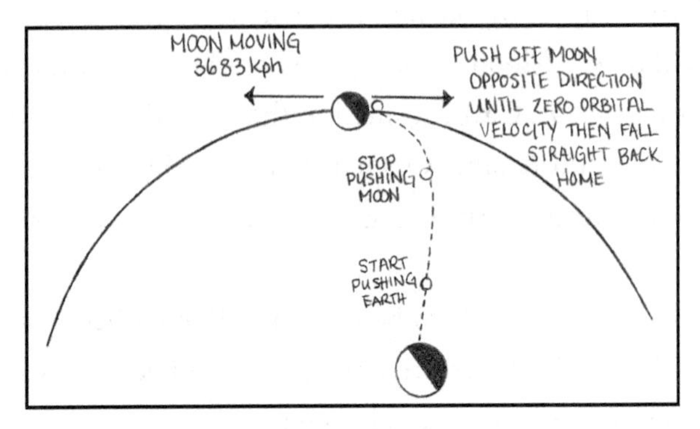

From the notebook of Dana Dash

Noah considered this and then said, "Ok, so what happens if we push off the Moon from here." Dana handed him the pen and he drew an arrow. "At a spot where we will push against the direction of the moon's orbit. Can we cancel the orbital velocity?"

"Right, that's what I was thinking," said Dana, "If we can kill the orbital velocity enough, then the Earth can pull us straight to it. Then all we need to do is switch to braking by pushing on the Earth at the right time and we won't hit the atmosphere too fast and burn up."

"So can we get enough push off the Moon to do that?"

"I don't know. I think the limit of the push we can get is the same as the escape velocity from the Moon. But maybe only half that because our neg-grav water tanks have to pull the pos-grav of us and the Beetle frame and windows along with them. But I don't know offhand what lunar escape velocity is, though the Moon's orbital velocity I know is 3683 kph."

Noah said, "I think we also need to be careful not to push off too much and end up orbiting Earth in the other direction. We need to know exactly when to close the shutters to the neg-grav and stop pushing the Moon."

"Or maybe we need to start pushing from the right place. If we start from this part of the Moon," she said, indicating the side away from the direction of the Moon's orbit, "we kill orbital velocity fastest. But if we start around here maybe," she continued, marking a point halfway between the trailing orbital edge and directly across from the Earth, "we kill the right side motion while getting extra push towards Earth to make the trip home faster."

"Ok, this isn't going to be easy to figure out," Noah said. "If we make it home alive I'm studying more math!"

"Agreed!" said Dana. "Do you think we can even figure it out?"

She tapped on her control panel to re-open the upper shutters, and looked at the Earth hanging there in the amazingly starry lunar sky. It was totally dark now, the Beetle having traveled mostly south, but also far enough east to speed up the slow-motion lunar sunset.

Noah looked skeptical and said, "There's no doubt we're both pretty bright, but I don't see us learning the necessary math in the few days we probably have to do this. Too bad we don't have Sir Issac Newton with us. Or one of those Math kids. If Toma and Tevi can design new cryptographic systems, I bet orbital equations are literally child's play to them."

"Yeah. I wish we could talk to them." Dana looked back up at the Earth. "Or anyone at all back on Earth really..." Then a thought came to her with a slow smile.

"What? You've got an idea. I can see it on your face! Please tell me we might actually have at least a chance to get back home," begged Noah.

"I do..." She pulled a cable out of her dongle drawer and looked around the cramped upper deck until she found what she was looking for. "And I think we might," she said, picking up the other ham radio handset that matched the one she had lost in the belly of a giant moon bug. One end of the cable plugged into the handset, and the other into a port on the Beetle brain. "I have software on this computer that will let me enhance faint radio signals from far away. Jayla showed it to me. If someone else is also running this kind of software, they might be able to hear us too. Even from Earth. But I am going to need a bigger antenna."

They were traveling rapidly above the surface again. Dana had shown Noah how to operate the controls, and as he flew the Beetle, she was moving about the tiny vehicle, working on her new project. Heading west, they had caught up with the blinding sunset again. Unfortunately, both to see where they were going and to use the pos-grav of the Beetle to accelerate them in that direction, some shutters towards the Sun needed to be at least partially open.

Dana faced away from the glare, trying to ignore the heat building up inside the cabin. She was using a tube cutter to cut spare lengths of aluminum tubing to exact measurements and fitting them together into a pattern with aluminum wire conduit connectors.

"So you really think this antenna will work to let us talk to people on Earth?" asked Noah.

"I think we have a good shot," said Dana. "It would definitely work better outside, though. The metal frame of the Beetle isn't close enough together to act as a Faraday cage but it will probably interfere some. If we still had a working space suit... but we don't, so no point in thinking about it. I need to make sure there's no contact between the antenna elements and any of these inner support struts."

"Yeah, you said before. Anyway, I'm optimistic."

"Me too. Jayla showed me how to clean the signal up with software on a computer. With the Beetle brains and this antenna, it should be enough to get a two-way voice link. Which is a good thing, because I haven't gotten around to learning Morse code[88] yet."

"Cool! You wanted me to let you know when we were close. The Earth looks like it's directly overhead now. Let me know what you think."

Dana looked up and said, "Yeah. Looks good. And it's getting pretty hot in here again. Slow us down and let's land. We don't necessarily have to be sitting still to make this work, but it might help."

---

[88] When signal is not good enough for voice, radio operators can send messages just by turning their radio broadast on and off in bursts that are short or long to make a code of dots and dashes.

Dana watched Noah work the controls but let him handle the landing by himself. When they were down, she took over the brains.

"I'm going to be using the brains with the radio, so it might take a minute to get the Beetle off the ground again if we need to. I need you to keep looking around and tell me the second you see anything move out there. We are *not* going to get captured by moon bugs again."

"No problem. I was noticing that some of our bacteria tanks are getting a little cloudy, and I want to take a look at them, but that can wait until we're done calling home."

"Oh, and can you get my telescope for me? It's down under the lower decking, next to where I was pulling the aluminum tube sections out for the antenna."

Noah retrieved her telescope, and she set it up right next to where she had positioned the antena, pointing straight up from the bottom of the Beetle to the top. The Earth above them was a crescent shape, not even half illuminated, and Dana got down under her telescope to look more closely at it.

"What are you looking for?" asked Noah.

"I'm trying to figure out when we will be halfway between New York and London. I could start transmitting a mayday signal, but even if I reach someone, no one is going to believe we're really up here, let alone help us with the math we need to get home."

"Right, that's why you wanted to talk to Jayla. You know she has a ham radio, and if we can convince her, she will go get the math kids. But wouldn't it be best if we were directly over New York instead of the Atlantic Ocean?"

"I can't call Jayla. It's not like a telephone. She has to be already actively listening, and at this distance, she has to be pointing her antenna at us while we're pointing ours at her. Also, the antenna elements have to be aligned the same way; a few degrees off and we might not be able to hear each other. Oh, and we have to be on the same channel."

"That is a lot of stuff to get right," said Noah, sounding worried.

"It is," said Dana. "But I happen to know that she is set up to communicate with her friend in London by moon bounce. I also know the frequency they like to use in the two-meter band. I built the antenna to optimize for that frequency. I know they planned to orient their elements north-south and I know they will only be able to talk for a short while when the Earth's rotation brings the Moon halfway between them — maybe an hour window. So I can predict the time pretty well."

"When will they start?"

"I can't see New York. It's late at night there. But I can see England, and based on its position from the day-night terminator, we probably got here at about the right time. Now, let's hope Jayla got this working, and that this is one of the nights she talks to her friend."

"That antenna seems to be a very specific shape," said Noah, "All the different pieces of pipe crossing the center support and I watched you measure each of them very carefully. How does it work?"

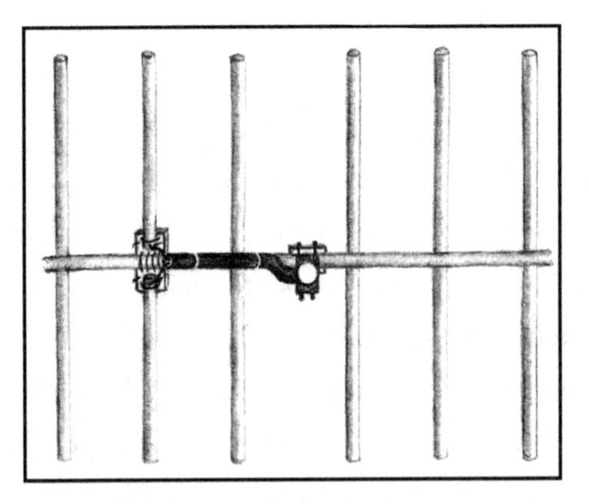

From the notebook of Dana Dash

"It's a two-meter SSB Yagi antenna. If you're really interested, I'll show you more specifically when we get home, but the frequency we want is close to two meters. That means that elements that are two meters long will pick up the signal, but the shorter elements you see on this antenna can also pick up the signal. They need to be of a harmonic length, so a half, or quarter, or eighth, whatever, of the wavelength. In this case, they're all half-sized, because that's the biggest I can fit in the Beetle. They're cut to specific lengths a little bit on either side of the frequency I know Jayla will use."

As she was talking, Dana had abandoned the telescope and was connecting wires. She unscrewed the whip antenna from the top of the hand unit and screwed in the connector for the antenna cable, and then she turned the ham radio on and sat down in front of the little laptop that was now both the control and communications center for the Beetle.

All she got was static.

CHAPTER 27 — MOON BOUNCE

As she listened to the static buzz, Dana realized she was holding her breath and let it out slowly. Just as slowly, she refilled her lungs, saying a silent prayer to the stars above to hear a voice — any voice — from Earth. Bosco was on the upper deck, and he scrambled over to put his head in her lap. Dana stroked his head and neck and ruffled his ears.

"CQ CQ CQ this is K8NNU Kilo Eight November November Uniform calling CQ, anyone there? Over" Jayla's voice came out of the computer speakers.

Dana's eyes popped open and she excitedly replied, "Jayla, it's Dana! Can you hear me?"

"I'm here too!" called Noah. After a couple seconds, he said," I don't think she can hear..."

"Hey Dana! I really want to know where you disappeared to, but..." She paused, then said, "I can hear you, Noah. But can we talk later? Right now I'm trying to connect with my friend Conrad on an EME bounce. Over."

"There is a delay for the signal to reach Earth," Dana explained to Noah. "We have to take turns talking and wait a couple seconds for replies. Go Jayla. Over."

"Oh, right!" said Noah and then remained quiet.

"Delay?" asked Jayla after a couple seconds. "Are you in England too? Over."

A new voice broke into the conversation, talking over Jayla. "20ADS Two Zero Alpha Delta Sierra here. Hey Jayla, who are you talking to? Over."

Dana and Noah waited quietly until Jayla responded with, "Hey

Conrad! I'm talking to my friends Dana and Noah. I thought they were in their lab — one of the other contestants here said they accidentally built their project too big to fit out the door. We all assumed they were doing their week-long experiment in there, but apparently they're in England too for some reason. Over."

"We aren't in England, Jayla," Dana said. "Noah and I are on the Moon and we need help. Over."

There was a pause, a few moments longer than the regular transmission delay. Then Jayla and Conrad both started speaking at once, making it difficult to untangle what each of them was saying, but the gist of it was that they thought it was a pretty lame joke. Jayla stopped talking before Conrad did and Dana could understand the last part of what he said. "...and her responses are following yours. Over."

Dana responded immediately, "Conrad, you're each hearing my responses to the other immediately because, as I said, I'm on the Moon. Over."

Again the two spoke over each other from where Dana and Noah were listening, their radio signals crossing as they reflected off the moon. But this time she managed to understand them.

Jayla said, "Her response to you had zero delay. Over."

Conrad said, "Her response to me had a 2.5 second delay. Over. But EME round trip is five seconds."

"Right," Dana replied, "because, you know, we're on the *Moon* here. Ov—"

"We really are," added Noah, "and we really do need your help. Over."

They waited quite a few seconds beyond the normal delay until, finally, Jayla and Conrad said the exact same thing. "*How* are you on the Moon? Over."

Dana took a deep breath, then gave the simplest explanation she could — Dr. Cavor's lab, the mystery machine, neg-grav, sabotage. She left out the parts about moon bugs or even mentioning Dr. Cavor at all. There would be time to give Jayla all the details later, but right now, they needed to focus on the emergency at hand. When they made it home — Dana stopped herself from thinking *if* they made it home — she could fill Jayla in on all the bizarrity over the past few days. But right now, Dana couldn't even think about it all. One thing at a time. Focus on the task at hand.

"...so, we need some help with the calculations to land correctly," Dana finished, hoping Jayla's practical side would beat her curiosity and see the need for more action, not more questions. Jayla asked just one.

"What you need is emergency math aid?"

Dana smiled. She knew she could count on Jayla. "Yeah, basically."

"Well, fortunately, we know some qualified math first responders. The twins won't believe this — everyone thought you were in your lab, and I hardly believe it myself — but I'll go get them. Over."

Conrad was still on the line. He and Noah talked back and forth for a few minutes. Suddenly Dana had a thought. She flipped through her lab notebook and found the paper with the names of all the kids who had attended the short-lived Moonbeam Preschool on the Cavor School campus.

"Hey, is your name Conrad R. Cartier?" she asked. "Over."

"Yeah — how did you know that? Over."

"A suspicion," she said. "It turns out we all went to the same preschool. Over."

"Hey!" said Noah. "That's a coincidence."

"You have no idea," said Dana, passing him the preschool class list. She hadn't shown him when she had discovered that all the other STEM contest finalists were on it, except for him. His eyebrows rose as he read down the list.

Conrad said, "Yeah, my father moved to London with me after my mother died..." He trailed off, forgetting to say "over."

Just then Jayla returned. "K8NNU here. I'm back with the twins. Over."

"Hey Dana! We hear you're on the Moon," said Toma. "Hey Noah! We hear you need a flight plan to get home," said Tevi.

Dana explained the nature of the math problem to the math kids, giving them her best calculations on the total mass of the Beetle with all passengers and equipment, and the amount of water that had been converted to neg-grav. It was clear that the twins didn't actually believe that Dana and Noah were on the Moon, but seemed willing to play along with the joke and excited to solve an interesting problem.

"We need to take a break from putting the final touch on our presentation anyway," said Tevi. "This will be a sort of fun problem to let us relax a bit," said Toma. "Over," added Tevi.

The twins headed off to crunch numbers. Dana told Jayla that she and Noah needed to decompress for a little while. Jayla and Conrad started chatting — they would only be able to talk to each other this way for a little while longer, but each would be able to adjust their antenna to continue to talk to Dana.

Toma and Tevi called back a couple of times to ask a few questions in between Dana and Noah's now familiar ordeal of eating, drinking, and navigating difficult bathroom procedures in very low gravity. Finally, they said they had a solution, but wanted to test their acceleration assumptions first.

"We need you to do a test take-off," said Toma. "And time it with a stopwatch," said Tevi. "Launch for as close to exactly one second as you can," said Toma. "And then turn off your thrust," said Tevi. "Then time exactly how long it takes you..." said Toma. "To touch down again," said Tevi.

Dana and Noah conducted the requested test and reported the results. The Beetle didn't climb very far, but it was enough for a bit of a bounce when they came back down.

"Ok," said Toma. "Now do it again for two seconds."

This time the rise and fall took a lot longer and the bounce was much more significant. When Dana reported the numbers, she said, "I don't think we want to go for three seconds. We might break something."

"It's ok," said Tevi. "We have what we need," said Toma. "The easiest way to do this is in four stages," said Toma. "One. We will need to negate your Earth orbital velocity," said Tevi. "Two. Then we will have you freefall back towards Earth," said Toma. "Three. Kick in thrust again, at the right time to slow down and reach the Earth at a reasonable speed," said Tevi.

Noah frowned. "That was three things."

"What is the fourth?" Dana asked.

"Zero. First you need to start at the right spot on the Moon," said Toma. "We need to get you to the lunar equator," said Tevi. "At the far lunar west, right between the near side and far side of the Moon," said Toma. "That's the easiest spot to calculate the launch to kill the orbital speed around the Earth that you have on the Moon," said Tevi.

"We're already pretty close to the equator," said Dana. "To have this conversation I got us as close to the center of the side of the Moon facing the Earth as possible to intercept Jayla and Conrad's moon bounce signals. So I figure I just head west, away from the Sun, until the Earth is exactly halfway visible over the horizon. That should put us in approximately the right spot."

"That's the right idea," said Toma, "But it will need to be a very close approximation," said Tevi. "You wouldn't happen to have a sextant with you?" asked Tevi. "Or be able to make one?" asked Toma.

"Hmmm, I —wait a minute! Would a medieval astrolabe do the trick?" Dana went and retrieved the bronze disc that Ms. Astrulabi had given her. "It has holes to sight through, there are gradations on the side to calculate sighting angles, and it can be hung from a string to make sure it's level."

"Sounds perfect!" exclaimed the twins in unison.

"So, will I be sighting on the North Star?" asked Dana.

"Yeah, sort of," said Toma, "we can use Polaris for one of the stars," said Tevi, "but it's not actually the north star for the moon," said Toma. "The Moon orbits over 5 degrees off the equatorial plane of the Earth," said Tevi, "so the Moon has a different north star," said Toma. "Omicron Draconis, actually," said Tevi, "but that's not important," said Toma, "because it'll be just over the horizon," said Tevi.

Dana went over a few stars that should be easy to spot in the sky and the twins calculated the angles she would be able to sight them at when the Beetle was in the correct position at the scheduled launch time. Then the twins gave her the timing for each of the other three stages of the journey.

Dana wrote down the info and said her goodbyes. Before she could cut the connection, Jayla came back on to say goodbye. She seemed very worried.

"It's going to be ok," Dana said with confidence she didn't feel. "The twins worked out the math. We're on our way home."

"Ok. Then... then I'll see you in a few days," Jayla said. She paused for a moment, and added, "Godspeed, Dana Dash."

As they began their journey, Dana explained to Noah how they would use the Earth and the stars to find the correct launch spot. "Navigation on the surface of the Moon is easier than on the Earth, because the Earth always stays in the same place in the lunar sky while you're standing still. The only way it changes is when you move."

"How does it do that?" asked Noah, "I thought everything in space was always moving around and orbiting other things."

"Oh, simple," Dana answered, "The Moon *is* orbiting the Earth, but it rotates once on its axis in exactly the same amount of time it takes to complete one trip around the Earth. So it's always showing the same face to the Earth. Looking from the Moon, this means that the Earth hangs there in the same spot in the sky while everything else in the cosmos goes spinning around."

Noah said, "Sort of like if we hold hands and spin around each other while skating, we look to each other like we aren't moving and the world spins around us."

"Right!" said Dana, "Except the Earth doesn't keep looking at the Moon. It's spinning faster than the lunar orbit. About twenty-eight times faster."

"Then is it a coincidence," asked Noah, "or does something make the

Moon's rotation match its orbital period?"

"It's not a coincidence. It's called a tidal lock. As the moon causes ocean tides on Earth, the Earth also exerts tidal forces on the Moon. Over time these tidal forces have changed the Moon's rotation to match its orbit."

"So how come the Earth's rotation doesn't change?"

"Oh, I think it does, but the Earth is bigger and it's taking a lot longer to slow down."

"How long?"

"Not sure I have it right, but I think I saw somewhere that it would be many many billions of years. But that will never actually happen because the Sun will expand into a red giant and eat the Earth and Moon in just a few billion years."

"Well, that's depressing," Noah said, "I guess we shouldn't be making any long-term plans..."

Dana laughed.

When they finally reached the spot specified for their launch they found they had some time to spare. They landed on the lunar surface and Dana went over the flight plan again while Noah spent the time playing with grow lights for the cyan-aid and taking some pH readings on the tanks — jotting things down in his log. "Our tanks are getting cloudier and I can't figure out why. Also, our batteries are getting lower now that we're on the dark side again," he said.

"Well, we'll be headed out shortly, and then the Sun should charge up our systems pretty quick. If we stay here, it won't be daytime again for a couple of weeks."

"Yeah," said Noah, "It's weird. I'm getting scared this time. On the way from Earth with Dr. Cavor here it was an adventure. I felt like nothing could go wrong with her in charge. And I never doubted that she could pilot us safely to the Moon and back. Even when the aliens had us locked up, she gave off this aura of invincibility. I knew things would be ok because

I could tell she believed it. But now it's like, if she can get killed, how do we have a chance? I mean, I like Toma and Tevi, and they're super smart, but can we really bet our lives on their math skills?"

Dana answered, "How do we have a choice? What else can we do, call our parents and convince them we're on the Moon and ask them to call NASA and launch a rescue mission? We only have so much food, and I'm pretty sure the cyan-aid system you built won't keep us breathing up here forever."

"I know. We have to launch and follow the time schedule they gave us. It's our only move but I'm still scared."

"Yeah, me too."

"Hey," Noah asked, "What are we going to do when we get home? We can't tell anyone about this, right? I mean, not anyone outside the school. Dr. Cavor said this technology needs to stay secret. But don't we need to tell people what happened to her? And what about those space bugs? The whole world should know about alien monsters on the moon, right?"

"I don't know. Let's think about that when the chances of us getting home alive look better. I'm thinking that, with our feet on the Earth again, it will be a lot easier to think about all this stuff and figure out what makes sense for us to do."

When they were ready, they did a pre-launch countdown. "T-minus ten, nine, eight, seven, six, five, four, three, two, one. Liftoff," said Dana. Then she opened the shutters, and crossed her trembling fingers. The Beetle bounced off the surface of the Moon and began to rise.

CHAPTER 28 — NOTHING LEFT, RIGHT?

The flight plan that the math genius twins had worked out was going to take almost a full day longer than the flight to the Moon had taken with Dr. Cavor at the helm, apparently winging it. *Add that to the list of amazing and inexplicable things that woman had been able to do,* Dana thought. She realized that this would make the day of their return the same day the STEM contest was being judged. She didn't see how they could win the contest at this point, even if they did manage to get back in time. But even though this had been such a big goal for her, now she couldn't find any spare mental space to care about a science fair contest. She didn't have anything left.

They used their meditation technique again to pass the time during the first leg of their journey. Dana set up a timer to activate one of the extra little servo motors that worked the shutters. When it was almost time to change over from being pushed by the Moon to being pulled by the Earth, the motor would run, winding up a piece of string on a spindle. The string was attached to Dana's wrist and would tug on her, until she woke up. Dana dozed dreamlessly.

When the tugging woke her up, she checked on Bosco and Chip before waking Noah. She noticed that the fox badly needed his bandage changed — he had managed to pull it about halfway off — but she didn't have any way to deal with that right now.

"Sorry buddy, You are going to have to wait until we get back home," she said to the disheveled canid. But she did clean out his cage, give him some food, and refill his water dish so he could drink now, while they were still under enough acceleration to keep some of the water in his dish. *Wait a minute — is that the same stone I swallowed? I thought the Moon bugs had taken it as a trophy. So the one currently rolling around in my digestive tract was just the Moon lady's rock?* She looked at the indigo rock in the water dish again, and picked it up to examine it. They really did look exactly alike. *What are the odds? Maybe,* she thought, *there are hundreds, thousands, millions of rocks like this whizzing through space all the time.* In that case, the odds would actually be pretty high. She put the rock back in Chip's dish and put the thought out of her overloaded mind.

Noah looked exhausted when she finally woke him. Dana didn't blame him. It had been one life-or-death crisis after another, with no guarantee of safe rest anytime soon. Noah blearily moved across the Beetle to the cyan-aid tanks to check on their oxygen levels.

"How's our air system doing?"

Noah tugged at his shirt. "Because of the way we took off, we didn't have much time to prepare. The lighting in our pod isn't equivalent to the sunlight the cyan-aid would've had at home. We've been using the supplementary lights that I would have run all night if we were doing this back on Earth. And the direct sunlight out here is too bright to use. That much ultraviolet light, which the Earth's atmosphere would normally filter out, might kill both my strains of bacteria."

"Both strains?" asked Dana. "Oh right. You said you had another oxygen producer besides cyan-aid."

"Right," said Noah. "Cyanobacteria are very efficient at photosynthesis, and I used the CRISPR to increase it, almost doubling oxygen production. I could've used pure cyan-aid for all our oxygen and we would've been fine. But I tried to also solve another problem."

"What problem?" asked Dana.

"Waste disposal," said Noah. "We've been using Dr. Cavor's meditation technique and not eating much, and no one wanted to use the waste collection devices unless absolutely necessary. So we haven't produced that much waste on this trip, but I loaded up half my oxygen farm tanks with a different kind of microbe, expecting to break down more. I wanted to surprise you and I thought it would impress the judges to have a more complete, self-sustaining biosphere system that included waste disposal."

"So these other microbes eat poop, instead of sunlight, to make our oxygen?" asked Dana.

"They do both," said Noah. "They're called cryptophytes. The 'crypto' part of their name is because they are strange creatures that both use photosynthesis and scavenge for other food sources. So I was able to use them in the oxygen farm and also for waste disposal.

"But something isn't working. The cryptophyte tanks are starting to look diseased." He moved aside so Dana could see. He was right; something was definitely wrong. Yesterday all the tanks had been a healthy murky blue-green, but now a lot of them had patches of brownish red spreading through the tank and the water was clouded slightly as if it had been contaminated. "I've tried both pumping in more CO_2 to the system and even adding some sugar I had which should have added a real energy kick to the wastewater, but so far there has been no obvious progress. In fact, what I did made the water cloudier." Dana moved closer to get a better look.

"We should be back on Earth in a few days, but..."

Noah sighed in agreement. "It's worrisome. I did one-third of the tanks with each type and one-third mixed, so the problem is affecting two-thirds of our oxygen production tanks."

"Should we try to contact Jayla again? It might be doable at changeover. Once we're in freefall we can point the Beetle any way we want to match Jayla's antenna alignment. Maybe she or Tevi and Toma would know what to do, or could do some research for us?" Dana suggested.

Noah shook his head, "They're good with engineering and math, not the feeding and care of algae. But there's someone who might be able to help — Prima, Lemma's partner. I remember her telling me about a science fair project she did where she grew algae in her fish tank to see how that changed the habitat and the lifespan of the fish."

"But she'll tell Lemma if we ask her for help. And there's no way Lemma doesn't rat us out. She'll spin this, and by the time we get back she'll have everyone thinking that everything that has happened is our fault — even that *we* killed Dr. Cavor. She'll probably even convince the world that we're responsible for the threat of alien bugs invading from the Moon."

"Dana!" Noah said. "This is more important than your feud with the Lei family."

"Fine-nuh," Dana said indignantly. But she knew he was right.

When it was time for the changeover to deceleration, Dana made sure the fox had finished eating and drinking, then warned Noah that there would be no feeling of gravity at all for a bit. She closed all the shutters — the next bit was tricky, as she explained to Noah.

"Ok, time to make a call back to Earth. Our antenna points to the top of the Beetle, and the natural tendency is for the landing plate to point to the gravity source, whether we're pushing or pulling. But we should be able to turn upside down and hold the position long enough to make a call.

"First, we need to do the hamster ball trick. Start at the bottom, then both of us quickly climb around the outside from here to there." They did this, then opened up the upper shutters that were now pointed at the Earth.

"There, now we sit here on the ceiling. Can you feel some gravity?"

"A little bit," said Noah.

"That's the Earth pulling on us, and we are dragging the inertial mass of the parts of the Beetle that the Earth can't see — particularly the shielded neg-grav and the landing plate which is now above us. As long as we sit here with the shutters open, we should be able to hold an exactly upside-down position." Dana got her telescope and checked the Earth far far below her crossed legs. "Ok, there's Italy... So it's late at night in New York, but hopefully Jayla will be listening for us."

"I hope she isn't asleep," said Noah.

Dana turned on the radio. "We need to orient our antenna elements properly. Here, crawl around this way. Follow me... good, that did it. "She took a breath and hoped. "Jayla, it's Dana. Are you there?... Hello?... Wake up Jayla! We need you."

After long seconds, a sleepy voice said, "K8NNU Kilo Eight November November Uniform. You know you're supposed to start with your callsign, Dana. Right? Over."

"Sorry. A little bit distracted up here, trying to stay alive and all. We're in trouble again. Our oxygen-producing system is having issues and we're worried we won't have enough air to get back. Noah wants to ask Prima some questions. Over."

"You want me to wake her up? Over."

"Yes please, if it's not too much trouble." she said sweetly, then abruptly loud, "Probably going to die up here! If we don't get some help soon it's all over! Over!"

"Sorry. I'm still half asleep. I'm on a handset, using my main radio in the lab as a repeater. I'll get her. Over."

Jayla disappeared for a while and returned with Prima. They heard Prima's voice. A soft, "Hello?"

"Hey, this is Dana. Noah's here too. The algae for our microbe-based oxygen farming system is sick — dying. Noah said you've done a similar project before."

"It could be because of a lack of real sunlight," Noah piped in. "I've been feeding in sugar and I supplied some extra CO_2, but it's not working."

"If it was because of the sunlight the water wouldn't be clouded," Dana pointed out.

They waited for a bit, then Prima said, "Well in case it's because of lack of sun, could you disconnect it from the pod and bring the tanks outside to soak up sunlight for a few hours in the morning?"

Noah and Dana exchanged awkward sideways glances.

"That's not possible considering our current situation," Dana explained. "You know how we've been absent the past few days. Well..." she paused for a long moment and then said it. "Didn't Jayla tell you? We are in outer space."

Her words were followed by silence on the other end. When Prima spoke again her voice was calm but she could hear the bulk of a million questions behind it.

"I would recommend trying yellow LED lights. Do you have any with you?"

"I'll check," Noah said. He left to rummage through supplies. "They're a bit old, might not even work but I can probably hook them up to something... need to make a closed circuit somehow." He continued to gather supplies and figure the lighting situation out while Dana and Prima talked.

"So... space." To her surprise she didn't sound completely opposed to the concept like Toma and Tevi, just extremely skeptical.

"It's a long story consisting of logistics that even I find unbelievable," she admitted. "I know it sounds crazy but we really need help right now, you don't have to believe us to do that."

Thankfully Prima's next question had nothing to do with their location. "So, you said the water is cloudy?"

"Yes."

"What types of algae are you using?"

"I believe the system has two kinds of microbes. Cyan-aid — I mean cyanobacteria — which eats light and turns CO_2 into O_2, and another type — I'm blanking on the name — crypto something..."

"Cryptophytes." Noah supplied.

"Right. They eat our wastes. The system is multifunctional. It refreshes air and disposes of waste," Dana explained.

Prima was silent on the other end of the radio connection for a few seconds. She could be heard shifting in her seat. "Do you think something went wrong with the waste disposal unit? That could explain the cloudy water."

Dana considered this theory. "It's possible. Noah?"

Noah nodded over his tiny LED light bulbs and glanced up at the cyan-aid tank. "We should sample the water before—" He was cut off by a loud ***BEEP BEEP*** and a flashing red light on the control panel of the oxygen farming system.

"What does that mean?" Dana checked the oxygen readings and gasped. "That was the alert I set up to warn of a dangerously low oxygen level range."

Noah's cheeks paled and he abandoned his light contraption on the floor. "What could be the cause of waste building up in the water? Some sort of blockage? Or malfunction of filtration? Prima, can you get to our lab? We have a few microbe tanks in there that we didn't end up using. You could experiment on them to try to figure out the problem." He told her the keypad combination to their lab door.

"Yeah —" She was cut off by another person entering the room.

A new voice spoke. "Who are you talking with?" Lemma demanded.

"Dana and Noah, they're in trouble and..."

"Dana and Noah? You mean our stuck-up competitors who can't solve a simple problem to save their lives? And you're helping them?"

"Hey, I told you. They're in trouble, like actual trouble." Prima went on to explain their predicament before Dana could tell her not to. Lemma burst out laughing.

"In *space?!* You actually believe that garbage!?"

Dana felt heat building in her chest, "You were the one who got us here! We know you and your sister were the ones who sabotaged our bio-sphere!" It was a bold accusation but it couldn't have been anyone else. The Leis wanted her gone from the competition; it wasn't that far-fetched.

"What? It wasn't me! There are plenty of people in this building who would be happy to take you out. Wait, what am I even arguing for? You're making all this up. Prima, this is ridiculous, hang up on them."

"No, I have to help them. They're saying that they're running out of oxygen." Dana heard her pick up the radio and start walking, while Lemma followed and continued to urge her to hang up.

Then the radio signal went to dead air.

"You really think it was Lemma who sabotaged us?" Noah wondered.

"I think that she'd do anything to get into that school and keep me out," Dana answered. Noah frowned but didn't say anything else.

They waited quietly for a number of minutes, occasionally trying a, "Hello? Anyone there?" into the radio. Finally the voice of Prima returned.

"Okay, I'm in your lab," Prima said. "I see the spare microbe tanks. One second." She set down the radio and all they could hear was Lemma continue to fuss behind loud fuzz for a minute before Prima returned to the conversation. "These microbes seem fine to me, but that doesn't mean the ones in your pod are. I found something else here too."

"Prima, don't. We're wasting time," Lemma commanded. "This is probably some sort of trick they think is funny."

"You don't get to order her around! And get out of our lab. We invited Prima in to help, not you," Dana shouted into the radio.

"I'm not ordering anyone, I'm stating my opinion. And my opinion happens to be that Noah and you don't deserve to be in this competition. Or at least you don't. You're trying to ride on Noah's coattails and are dragging him down. Noah should get wise and lose the dead weight."

Dana turned red and opened her mouth to let loose on Lemma when Noah grabbed the radio from her and held up a hand. "Prima, please, what did you find?"

"There are two scales and sets of weights here. One is glued to the bottom of the table. What is that about? And why are the labels on the upside-down weights mirror images of the others? Did you have a custom set made as some sort of joke?"

Noah started to explain how when the gravity-reversing machine changed the weights to neg-grav, the labels got reversed. But Lemma interrupted him. "I really don't see why this nonsense is relevant."

But then Prima gave out a little yell, and Dana heard the familiar sound of a neg-grav metal weight, accidentally dropped upward, hitting the ceiling of the lab. There was a long pause and then Prima yelled, "Wow! You're really in outer space?!?"

"That's what we've been *telling* you," said Dana.

"Oh my gosh! So if this machine reverses the labels too, I believe I've figured out your oxygen problem. How did you set up your system?"

"Dana built the capsule and I engineered the microbes."

"And the water tanks. They're the anti-gravity mass which you use to lift you?" asked Prima. Dana marveled at how quickly she was absorbing and deducing new information.

"That's right," said Noah. "We needed to move our project out through the retractable roof. It didn't fit through the door, so we — "

"Right," interrupted Prima, "And at what point in your build did the water get converted?"

Noah opened his mouth to reply, then snapped it shut again as if he'd been struck with an epiphany. "I engineered the microbes, then added them to the tanks..." He met Dana's eyes in sudden realization. "That's where we went wrong!"

"Yeah," Prima agreed. "I knew it. You didn't convert the water until after you'd already added the microbes!"

"Ok," said Dana, "so it's neg-grav cyan-aid. Does that make the bacteria sick somehow?"

Noah said, "Not the neg-grav part. But it also flipped the microbes right-left in spatial dimensions! Just like it reversed the labels on the weights."

Noah was doing the small-confined-spacecraft-environment equivalent of pacing back and forth: rocking back and forth from his knees to his heels. "That means that we accidentally changed the microbes to have a different chirality!"[89]

"They can't make oxygen because they're mixed up about right and left?" asked Dana.

"They can't eat the waste material," said Prima, "or any of the sugar you gave them. It's hanging around and blocking their light."

Noah said, "Organic molecules can be thought of as being left-handed or right-handed. Not really handed, of course, but the same molecular

---

[89] *Chirality* is a fancy word for saying a chemical compound can have a mirror image.

formula can be built two different ways that are mirror images of each other. Like this." He sketched some molecular diagrams in his lab notebook.

"So, left-handed algae will only eat left-handed food. If you flip the algae and make it right-handed, it can't eat the left-handed food anymore. And our poop is left-handed. Pretty much everything on Earth is. So our waste is building up, uneaten, and clouding most of our oxygen farm tanks." Noah sounded elated at their having figured out the problem.

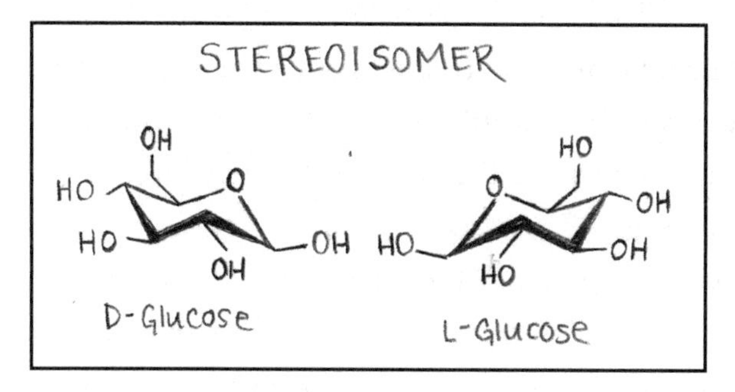

From the notebook of Noah Knight

"Ok, right," said Dana. "Now I remember about mirrored chemical structures called stereoisomers[90] from chemistry. So we know what's wrong. But why are you so happy? If we don't figure something out, soon we're going to suffocate."

The excited look on Noah's face vanished with the reminder that this problem was not academic but, in fact, quite critical to their very survival. They had only figured out what went wrong, not how to fix it.

Unless they could make the air system work again, and soon, they were going to die gasping on stale air.

---

[90] People often think that the *stereo* part of *stereoisomer* means two ways (right and left) like a stereo has two speakers. But it's actually from the Greek meaning "three-dimensional," and *iso* means "the same," so it is the same molecule but can be different in three-dimensional construction.

CHAPTER 29 — BREATHING EASY

Noah stared at the bacteria tanks, his brow deeply furrowed as if he could glare at the problem until it went away.

Dana could hear Lemma faintly in the radio background berating Prima. Lemma was trying to order her to leave the lab and come back to the dorm. Dana heard her say, "Honestly. It's just a big joke they're playing on us. How could they have remembered to bring water but not have enough oxygen?"

Noah said, "Thanks for all your help Prima, we may call you back later." He hit the red button on the screen to stop the radio program and turned to Dana. "I know what we have to do, but you're not going to like it."

"Ok. What?"

"We have to filter the water out of half the tanks, and start again from scratch."

"Just the cyan-aid," she guessed.

"Yeah. Fortunately I have a couple tanks without mixed bacteria types. We can store our wastes until we get home. The cryptophytes were really just for showing off when we were trying to win a stupid science contest. They're not critical to staying alive."

"So, rebuild our oxygen farm in space?"

She rubbed her temple and shut her eyes for a second, taking a deep breath. "Ok. Let's do it." Noah nodded and turned to the tanks.

They pointed the Beetle's landing gear at Earth and started dropping towards home. Then they got to the business of fixing the air system. They filtered the polluted tanks, one at a time, into a storage bin that had previously held tools. The low gravity created by Earth's greater pull on the contents of the Beetle than the spacecraft as a whole caused the water to misbehave badly. As they filtered the water, it climbed the sides of the bins and spilled all over the place. By the end, a great deal of foul-smelling sewage water was pooled in the airlock. They replaced lost water from their drinking water reserves and Noah stirred in some unclouded cyan-aid to start each new batch growing.

"Is it working?" Dana asked, wiping beads of sweat from her forehead. Noah nodded, but it didn't look like a certain yes.

"It should, but —" He stopped mid-sentence to stare at the oxygen gauge which had lost another light.

"It's still dropping," Dana said. "The new cyan-aid tanks must not be active yet. They're not making enough oxygen."

"It'll be a while before they can produce enough to start making any real difference. I increased their metabolism when I increased oxygen production, but even with the faster doubling rate, and feeding the new cyan-aid a lot of light, it will be at least a day. Maybe longer."

"If they don't start soon... "

Their eyes met in silence. They'd done everything they could, but it wasn't going to be enough. This one stupid mistake was going to kill them, and no one would even know. Would anyone even find their spacecraft? Would anyone track it back to the Cavor School STEM contest? And if so, what would they think? Some genius kids accidentally ended up in space? Probably not...

Dana already felt like she couldn't breathe. The air was heavy in her lungs and the spacecraft felt even more small and cramped. They had gotten Dr. Cavor killed. Maybe it was only fair that they got themselves killed too.

When Noah spoke again it was in a strange tone of voice. "There's only one other thing I can think of that might help."

Dana looked at Noah, but he wasn't meeting her eyes. She followed his gaze, and realized what he had in mind as her eyes fell on Bosco, curled up asleep on the upper deck.

"You can't be thinking..." she started.

Noah interrupted, saying, "I know this is hard, but think it through! We have an extra dog and a fox on board. I'm not sure how much air that fox uses, but that dog is as big as one of us."

"Bosco and Chip! Say their names!" Dana was yelling now. "They are not just 'that dog' and 'that fox.' They are part of our family. We have a duty of care. Dashes don't abandon their commitments!" That last part was her father talking, but it was a truth she had long ago internalized as her own.

Bosco was awake now and looked concerned to see the two of them squaring off against each other. Chip was agitated and scratching in his water dish full of stones.

Noah looked somewhat cowed by the yelling but was still determined. "Ok. Ok. I hear you. But I need to state this clearly. I don't think what we did is going to work in time to save us. Not any of us. And if it were just you and me, we would consume less than two-thirds as much oxygen. Our odds of surviving would be considerably better."

Dana was no longer shouting, but her voice was bitter and cutting. "Oh, and you're going to kill them are you? How? You can't put them out the airlock without losing a bunch more air. Stab? Strangle? How are you planning to kill my dog who is basically like a furry brother to me?"

Noah said quietly, "Survival isn't always easy. And scientists sometimes have to kill lab animals..."

"No," said Dana. "It's not going to happen. We live or die together. If you're going to kill them, then you better be prepared to kill me too."

"Wow," said Noah. "I hope you know I would never even consider..." He fell silent.

"Right," said Dana. "And that's how I feel about Bosco and Chip. I would never even consider it. They're family. They're my responsibility. And you are too."

"I just don't want to die," said Noah. His face was full of fear.

"I'm sorry," Dana said softly.

"It's not your fault. I should be saying sorry. It was my oxygen system that failed."

"It's not your fault either," she said.

"But if I'd thought about the results of putting microbes through the gravity-reversing machine... I should have known..."

Dana put a hand up to stop him. "We've both made mistakes. I'm not the bio-geek, but I do know chemistry, and I didn't make the connection between reversed labels on the weights and molecular stereoisomers. And we wouldn't be here at all if I had measured the door before building the Beetle's frame." She started to think about all the questionable things she had done to lead them here. Maybe Principal Peters was right. Maybe she was a menace. "If you need to blame yourself, blame me too. We can be at fault together."

They turned to look at the tank. It had a tiny scale on it with five oxygen indicator lights. As they watched, the second-to-last light went out, leaving only one blinking. Dana took Noah's hand and squeezed his fingers hard, and he squeezed back. There was no way the second set of tanks was going to kick in fast enough to save them.

They did the only thing they could think to do to conserve oxygen: they curled up next to each other and Bosco, and went into Dr. Cavor's meditative state. Dana closed her eyes and tried to quiet her thoughts, letting them melt away behind a deep layer of fog. She breathed in slowly, and like before, she was soon fading. Everything in her head was a swirl of pastel light and ice. She could feel Noah next to her, and faintly hear Bosco breathing, but that was it.

The *beep-beep-beep* of the oxygen sensor warning turned into a monotone hum in her head. Somewhere, far away, she knew she was dying. Meditation wasn't going to be enough to save them. She had hours — or maybe only minutes — to live. The alarm would go off, days from now, when they should be reaching Earth, but no one would wake up to answer it. Maybe the Math kids hadn't even gotten it right. Maybe they would hit the atmosphere unimaginably fast and burn to ash in an instant like a shooting star. Would anyone look up and make a wish?

She knew she was doing Dr. Cavor's meditation right, but one nagging thought stopped her from slipping completely into a trance. She could hear

Lemma's annoying voice saying over and over *How could they have remembered to bring water but not have enough oxygen?* Was she going to die because she couldn't shut up that stupid voice enough to meditate deeply? Deeper. Deeper. Fading. And then Lemma's voice again, *How could they have remembered to bring water but not have enough oxygen?*

Dana jolted fully awake. If Lemma was there right now she would have kissed her! She was going to live! They were all going to live!

Noah was deep in a trance, and he remained so as Dana scrambled to the spare parts bin and started to build. The air was stale and choked her, and her whole body felt light and tingly from lack of oxygen, but the thrill of having a working plan helped keep her focused. Soon she had assembled a device of electrodes, rubber tubing, and a plastic water tank with a divider in it. She connected the power from the solar cells, stuck a tube in her mouth, and sucked on the gas it produced. Soon, she started to feel a lot less lightheaded.

As her body normalized, she reflected on how Lemma's words had saved her. If you have tanks of water, of course you have oxygen — a lot of it. *The molecular formula for water is H_2O. Two atoms of hydrogen and one atom of oxygen for each molecule of water. And since liquid is a more condensed form than gas, the oxygen from each liter of water — our beautiful, beautiful water — ends up being over five hundred liters of pure, breathable O_2 gas. Enough for one person to breathe for a whole day. The hydrogen just needs to be separated from the oxygen. All you need to do that is an electrical current to break the bonds, pull the hydrogen and oxygen ions apart, and then recombine them into H_2 and O_2 gas.* And with the full sunshine available in outer space — no clouds, or even atmosphere, to reduce the output of their solar cells — Dana could easily make enough electric current to produce as much oxygen as she wanted. She only needed to find space to put the excess hydrogen, and the vacuum outside the Beetle's windows had a lot of available space.[91]

---

[91] That's why they call it "space," after all.

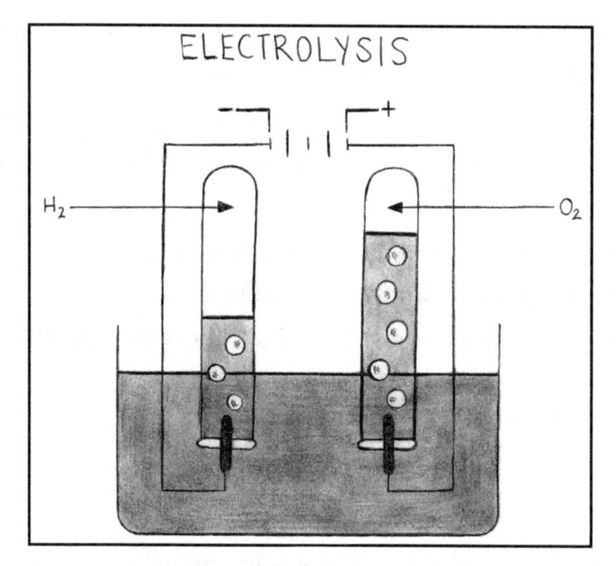

From the notebook of Dana Dash

She did a few rough calculations and added a couple things to her electrolysis device. The oxygen flowed directly into the spacecraft cabin. For the hydrogen, she used a spare pressure control valve with a screw adjustment to set the right cracking pressure. This would release the hydrogen into space before it built up too much. She poked a hole through the little meteor-hole patch and fed the hydrogen tube through it, gluing the patch back into place.

She watched to make sure the valve was working, then went back to meditating. But the air seemed to choke her more than ever, even as the oxygen gauge showed it was rising back towards normal. Something was still wrong.

The carbon dioxide! Her setup was producing oxygen, but unlike Noah's algae farm, it wasn't also consuming their exhaled CO_2. Noah had told her that a person could die from too much CO_2 even while getting plenty of oxygen! But how could she get rid of all this gas that the plants loved, but was killing them?

She sucked a breath straight from the oxygen tube to clear her head, but she found too much straight oxygen also made her lightheaded and dizzy. So she alternated pure O_2 from the tube with breaths from the too-much-CO_2-cabin-air. This did the trick, but would require constant attention and wasn't a good long-term solution considering they would have to be doing this for days. Not to mention that Bosco and Chip couldn't be trained to do this trick.

Noah! He was still in a trance, breathing increasingly poisonous air. She shook him repeatedly until he started to wake up.

He coughed and said something like, "shooomerzle, whatshuh? Are we dead yet?"

"Breathe this," said Dana, handing him the oxygen tube.

Puzzled, Noah took the tube, and after one breath he began to suck on it greedily.

"Oh yeahhhhh... That's the good stuff!" he said.

"Pure oxygen," said Dana. "Something Lemma said got me thinking and I built an electrolysis rig to convert water into hydrogen and oxygen."

"Then you saved us!" Noah was elated.

"Not quite yet," said Dana. "We can't really breathe pure oxygen, and I still don't know how to get rid of the carbon dioxide we're exhaling. I've been alternating between light-headed on pure O_2 and choking on CO_2, trying to keep the right balance, but it's not easy."

Noah handed Dana the oxygen tube.

"Hmmm." He thought for a minute and then said, "Oh! One of those is actually really easy! Getting too much oxygen is only a problem if too much gets into your blood. And how much gets into your blood depends on the air pressure around us."

"Oh right!" Dana was excited. "Mountain climbers have to use pure oxygen above a certain altitude because the air won't push enough oxygen into their blood at that altitude."

"Right," said Noah. "So if we lower the air pressure, we could breathe straight off your oxygen machine indefinitely with no problem."

Dana handed the oxygen tube back to Noah.

"Easy enough to lower the air pressure in here," said Dana. "We open the door a little. I was noticing that I already had some pressure buildup from my oxygen flow, and was thinking that I'd have to crack the airlock door with the crowbar and reduce it at some point. We just need to do that more often to keep the pressure lower."

"Ok," Noah said, "So, we can breathe pure O_2 straight from a tube and not worry about the CO_2. Are we saved?"

Bosco, who had been panting faster and louder, now coughed and growled.

"Oh!" said Dana. "Bosco can't breathe through a tube. Chip is in an aquarium. I can mostly seal him off with duct tape and feed in an oxygen tube, and his exhaled CO_2 will be pushed out by the oxygen flow. But what about Bosco? He's already having problems breathing."

They sat and thought for a couple minutes, passing the oxygen back and forth. Then Noah's eyes lit up.

"How about the space suit? Could we make a breathing mask out of the fabric?"

"Maybe!" exclaimed Dana. "If we can get him in it and keep him there, it will be just like Chip's aquarium, O_2 in pushing CO_2 out."

Dana fetched the space suit from the plastic bag, trying to ignore the stench of green-cheese-moon-fungus still clinging to it. Bosco objected only a little to having his head forced into one leg of the space suit pants and having it zip tied around his neck. He tried to move backwards a few times to escape, but kept running into walls. Dana stroked his back saying soothing things, and eventually, he lay down, resigned that this was his life now. Chip didn't seem to notice any change in his aquarium at all.

The children took turns sleep-meditating, while the other stayed awake to watch Bosco, choke down a food-like bar from the rations, and periodically crack the door when the pressure got too high. Time passed like this, marked in cycles of wake-sleep-wake-sleep-wake-sleep. When the alarm sounded again, indicating that it was time to put on the brakes, Dana made the necessary adjustments, and didn't bother to tell Noah until it was his wake-time, and her sleep-time.

Dana was mastering meditation. Exhaustion, constant fear, and stress from each new problem had overwhelmed any discomfort from the breathing contraption. During her wake-time, she couldn't hide from her thoughts: all the strange and horrible things she had witnessed. A machine that reversed gravity. A man who breathed lizards. A garbage creature controlling whatever it touched. A well-mannered but horrific moon-queen-hydra-slug-monster. Dr. Cavor, her mentor-alien?-suicide-bomber, dying in a war Dana didn't understand. At the end of each of her wake-times, she couldn't wait to release her churning thoughts.

For each cycle of sleep-time, her last thought, before surrendering to the sweet blankness, was relief that while they might all still be going to die, at least things couldn't get much weirder.

CHAPTER 30 — THINGS GET MUCH WEIRDER

Dana knew she was asleep and dreaming but everything felt real and vivid. She was in a large clear box and could hear a constant hissing noise. There were torn gauze bandages all over the floor in front of her, with colorful rocks scattered among them. She reached out, but couldn't pick them up with her paws.

Paws? She looked down at her body, covered with red fur. She she was a fox! Ah, now it made sense. She was in Chip's aquarium. Through the glass she could see Noah in the airlock, holding one end of the crowbar, releasing some air pressure into space. There was her own body — where she belonged — asleep in a hammock.

Dana woke up with a start and sat up. Noah was pulling the crowbar out of the door. From the fox's aquarium, Chip was looking right at her, as she knew he would be. His bandages were removed and he looked at her with one eye. The right eye, which should have still been a messy scar, was healed over perfectly — smooth fur, like there had never been an eye there at all. She thought she saw a sparkle for a second — a twinkle in his non-eye.

She removed her oxygen tube, got out of the hammock, and took the crowbar from Noah. Then she pointed it at the fox's cage and said, "Ok, What's going on here?"

Noah looked and said, "Chip knocked his bandages off? But he looks ok."

"Yes he does. Too ok. Look at his eye!"

"Huh. How could he heal like that..."

"No idea. I've seen him with the bandages off before. His right eye was an ugly hole."

"Ok," said Noah. "That's admittedly a little weird, but only a little, compared to some of what we've just been through. Any idea what could have healed him so perfectly?"

"Maybe," said Dana. She knew it was more than healing but she didn't want to tell Noah that she had been seeing through the fox's eyes (eye?) before she woke up. She approached the cage and confirmed what she already knew. Chip's water dish was tipped over, and colorful rocks and torn bandages were scattered all over his cage. The little indigo meteorite was nowhere to be seen.

As she turned back to Noah, saying, "I think it has something to do with that little meteorite I dropped in his water dish..." the crowbar in her hand bumped one of the aluminum struts, ringing the Beetle's frame like a bell.

A voice — *a* voice, not Noah's voice — spoke.

"Could you gently put down that crowbar before you bump something else?"

Dana spun around. "Did you hear that?" she asked.

"You mean when you turned the ship into Big Ben?" asked Noah, rubbing his ear. "How could I miss it?"

He hadn't heard it.

"I do not want to alarm you, but the atmosphere in this capsule now contains a large quantity of uncombined hydrogen and oxygen gases."

Dana spun around again and stared at the fox in his aquarium. She raised an eyebrow, and Chip nodded. He nodded at her!

"It would not take much for it to recombine in a huge explosion... one tiny little spark... Go, see for yourself."

She gingerly laid the crowbar on the floor, and went to look at the electrolysis device. Sure enough, the hydrogen side was full and the gas was bubbling up out of its container and into the cabin of the Beetle. The exhaust valve wasn't opening up. Why? *Oh! Duh! We lowered the pressure of the whole cabin so we could breathe pure oxygen, but I didn't adjust the pressure screw on this valve to compensate for the change. Of course the hydrogen has been backing up into the cabin rather than being released into space.* She pulled a screwdriver out of her tool kit and adjusted the valve, careful to first touch the metal of the tool and the screw with her fingers first to distribute any difference in electrical charge[92] without a spark.

"We need to be super super careful not to make a spark," she said.

"Why?" said Noah. "What's wrong?"

"The escape valve for the hydrogen gas wasn't working right. We've been filling the Beetle with both oxygen and hydrogen. We almost *Hindenburg*ed[93] ourselves. Give a big thank-you to whatever put the idea in my head to check on it."

"You are most welcome," said the fox that only Dana could hear.

"Wow," said Noah, a little freaked out. "That's scary."

Dana checked the flight clock. "We have time for one more short shift before we're close enough to Earth to see if the math twins' math has saved us. Your turn to sleep."

As soon as he was out, she asked, "Ok, what in the space squid is going on here? Who — *what* — are you??"

"I believe you call me Chip, or do you prefer Alien Bralien Chipalien?"

"How am I hearing you? How do you know all this stuff? Your name, the mistake with the hydrogen valve, how the servo motors work, any of that?"

"The stone healed this body. Bonded with me when it touched the wounds, the blood. I don't understand it all. I feel I am a small piece of what I once was. Most of what I know is from you. You are, to some extent, talking to yourself. There is more from the stone that is not you, but my

[92] Static electricity is real electricity. That shock you sometimes get when you slide around on carpet in your socks? Real enough electricity to blow up, say, a hydrogen-filled German airship. Or the Beetle.

[93] The *Hindenburg* was an airship, a type of flying, rigid-framed balloon, that carried passengers. It was filled with hydrogen, and caught fire, and a bunch of people died. This was in in 1937 before spaceships.

name, the hydrogen valve, the motors — these are all things that you knew, somewhere in your mind."

"And that connection? Why are you in my head? Why can I hear you, but Noah can't?" But then she knew.

"Correct. You swallowed the other stone. But not any other stone. Those two stones are the same. Echoes of each other."

"What do you mean 'echoes'? How can they be the same stone?"

"That I do not know. I know the stone and I are part of you. You and the stone are part of me."

"And what if I don't want to be part of anyone else?"

"The stone is not inside you the way it is inside me. It is merely passing through the center of your torus.[94] It is inside your electrical field but not really inside you."

"Wait... you mean...?"

Noah was asleep, and it had been a while since Dana had taken care of certain bodily functions. So Dana went into the airlock and made use of one of the solid waste disposal kits they had created. Then, green and gross as it was,[95] she poked through what she had produced and found the stone. It had, in fact, passed through her body.

"Are you still there?" she asked.

There was no response.

She retrieved the stone from the bag and used the available wipes and some water to clean it vigorously until she was sure all waste material had been removed. Then she held it tightly in her hand.

"Hello? Are you there?"

Still no response.

---

[94] A *torus* is a donut shape. If you think about a human being's digestive tract from one end to the other as the hole in the middle, a human being is a very lumpy, misshapen, donut. Things that you swallow are not actually "inside" your body, in the same way that putting your finger through a donut hole isn't the same as putting your finger inside the donut.

[95] Eww. Don't do this. Unless it's to find magic-talking-fox-mind-meld-space-stones. Then it's ok.

She sighed. Repeat the experiment. No matter how gross it is. "The things I'm willing to do for science...."

She wiped the already spotlessly clean stone extra vigorously one more time. Then she counted, "One... two... three..." and popped the stone into her mouth, like the world's weirdest and possibly most disgusting piece of candy.

"***Ewwww! That is truly disgusting,***" Chip laughed in her head. "***I am having second thoughts about being bonded with you.***"

She spat out the stone and said "Oh, shut up!"

And, of course, he immediately did.

She put the stone in her pocket. If she needed to talk to him, she knew how it worked. Right now, she needed some time, alone, to think. From the aquarium, Chip stared at her as she settled into position next to Bosco, watching the pressure gauge and petting the poor, pants-headed pooch.

When Dana's alarm sounded, she woke Noah and he moved to the controls. He took the oxygen tube out of his mouth and said, "What now?"

"The alarm was set for five minutes before Tevi and Toma's flight plan said we should take a peek and play it by ear. A countdown timer should be running in the top left corner of your screen. Watch when it gets to zero and switch to freefall with the outer shutters open."

A noise from Chip's cage got Dana's attention. He was pawing frantically at the glass.

"Belay that order," said Dana. She slipped the stone in her mouth.

"Are you certain those little servo motors won't make a spark?"

Dana spit the stone out. "The servo motors might make a spark," she said. "I'll work the shutters in the airlock by hand if needed. You'll only open the outer ones. Their servos are outside. Tell me when that timer hits zero."

She climbed down to the lower deck and carefully detached the springs that held the shutters in default positions. When Noah alerted her it was time, she said, "Ok, fingers crossed the twins got it right."

"You don't trust their math?" asked Noah.

"I trust them to have done the math right," said Dana. "I just hope we got them the right info. They had us run a test, but we could only estimate the different masses and how much actually gets screened off by the gravity shutters. Ok, ready? Time to see where we are!"

The shutters opened, and the Earth loomed beneath them. It was still growing larger as she watched, so she left the brakes on for the moment.

"We're so close." Noah was at the window, his face pressed up against the plexiglass so his breath fogged up the view. "What's the first thing you're going to do?"

"Talk to the teachers about what happened with Dr. Cavor. Check for any needed repairs on the Beetle," Dana said dutifully. "Win the contest? We may get back in time for judging, if we haven't been disqualified for being missing this whole week."

Noah snorted, "Really? I'm going to take a long shower. Then contemplate the meaning of existence for at least a few hours."

Dana laughed. "You do that. And I'll do *all* the work on the ship."

"We already know it works. We spent *how* much time in here? And we took it to the actual Moon! We should get, like, tons of extra points for that. Not to mention we could have died a whole bunch of times. How much is each almost-dying-but-actually-surviving experience worth, do you think?" he asked. Dana rolled her eyes. Noah feinted shock. "It's nothing to eyeroll. Think of the trauma! We should sue!"

"Who? There's no one to file a lawsuit against. This was all us. Except maybe Lemma or Lambda Lei."

"You don't seriously still think the Leis had anything to do with this?"

"I'm just saying, she —"

"*Dana*," Noah said, souding just a touch exasperated. "The Leis might not be good people, but she didn't sell us a one-way ticket to the Moon or bribe space bugs into attacking us. Okay?"

Dana looked at the still-growing Earth below them. "I just need to figure out when we ease off on the brakes here. If the twins calculated everything right, we should be going slow enough to float down through the atmosphere without..."

The Beetle started to vibrate. *HmmmMMMMMM*

"What is that?" asked Noah.

"I don't know, but it can't be good," said Dana.

The vibrations increased; it was no longer a hum but a roar. Dana started to feel heat radiating off the floor and walls. The lower shutters were starting to glow, and the roaring was becoming deafening.

"We got the math wrong," Dana said. "We're coming in too fast. We're burning up. "Her voice was weary. She knew she should be screaming, but she felt like all the worrying and screaming circuits in her brain had finally overloaded. She would calmly await the end of her world...

...and then the gears in her brain re-engaged and she snapped back to herself. Calmly? Nope! She refused to burn up. Not after coming this far. She hadn't tried everything, and she wasn't giving up yet. *What are the threats? Entering the atmosphere too fast might create enough heat to burn us up. Hydrogen gas is still intermixing with the oxygen and carbon dioxide in the Beetle. If fire breaches the exterior at all we might explode. Think fast!*

"Noah! Close the lower shutters. We need to make vents. "She pointed. "There! Where the legs connect to the frame, pull up the decking and pull away the rubber seal!"

Dana dove for one of the five corners and ripped at the rubber gasket. The air inside of the Beetle, at a higher pressure than the low-atmosphere air outside, vented out in a strong jet of hydrogen and oxygen and ignited with the friction, creating a booster rocket to help them decelerate.

Dana started tearing at a second gasket. Right now, the base plate was taking the brunt of the atmospheric heat and friction. If it became unbalanced, and tipped, the Beetle would be torn apart.

As she tugged her second gasket, she saw that not only was Noah working on his second gasket, but Bosco had torn off his breathing-pants-rig and was pulling at the final gasket with his teeth. In almost no time at all, jets were firing out of each corner of the base. The Beetle had improvised retrorockets.[96] They were slowing their fall.

The force of the deceleration was like a falling elevator suddenly stopping — and they were all pressed hard against the floor. Barely able to move, Dana felt the floor continue to get hotter and hotter. Her ears roared, and the light inside the Beetle took on the demonic hue of the flames dancing up from the base plate below.

Then, slowly, things began to improve. The roar dwindled to a fierce purr. Deceleration pressure decreased and Dana was able to peel herself off the floor. Her head was spinning but she needed to reconnect the springs on the airlock shutters to control their landing. It was hot in the airlock, and it was only the thinness of the air that allowed her to keep going. In thicker air the heat might have cooked her. Her fingers burned as she reconnected the shutters and her head was swimming from lack of oxygen.

They had vented most of their air pressure and would need to drop to where the air was thicker. With the springs reconnected, and the shutters pulled closed, the Beetle was falling. But enough of their speed from re-entry had been slowed so that they were now only falling as fast as the increasing air resistance would allow. No more flames came from below, and the heat, while almost unbearable, was no longer getting worse.

But she knew they were not out of danger yet. If she passed out from lack of oxygen now, there was no guarantee she would wake up in time to land the Beetle safely. As she climbed back up to the controls, a red fog started to blur the edges of her vision. She sucked on the oxygen tube, but it didn't seem to help. Then she remembered what Noah had said about pressure and getting oxygen into the blood. They had originally lowered the pressure so they could breathe pure oxygen safely. But now the pressure

---

[96] A *retrorocket* (short for retrograde rocket) is a rocket engine that provides thrust opposite the motion of a vehicle, causing it to slow down. It's basically like having your brakes be fire and wind. Cool. Scary.

was even lower — so low that even with breathing pure oxygen, not enough was getting into her blood. But how could she increase the pressure inside her lungs before she fainted? Her vision began to telescope to black.

As her eyes blurred and she felt herself start to slip out of consciousness, she had a last, desperate idea. She inhaled a deep breath from the oxygen tube, filling her lungs like a balloon. Then held her nose and kept her lips tightly closed while she breathed out — hard. Her diaphragm muscles tightened, and the pressure in her lungs increased. Her vision returned, the fuzziness and red ebbing to her periphery. It was a complicated process to go through with each breath, but it kept her conscious, albeit dizzy, and she was able to work the controls to open the lower outside shutters. The Earth was now able to "see" the pos-grav inside the Beetle, and pulled the craft down like a regular object falling out of the sky. Soon, they were down to an atmospheric pressure where she no longer had to self-pressurize her lungs to breathe.

She closed the lower shutters and halted their descent. With the Beetle stabilized at an altitude a few kilometers higher than Mount Everest, she was able to help the others. She found Bosco's pants-based breathing mask and stuffed him back into it. Noah had lost his breathing mask when he passed out, so she put it back on his face. Moments after the oxygen hit his lungs, he regained consciousness.

"Uggghhhh..." He started pulling himself off the floor. "What happened? I feel like someone hit me."

"When we vented atmosphere, you passed out from lack of air pressure."

"And you didn't pass out? You aren't some sort of alien super woman that doesn't need air, like Dr. Cavor, are you?"

Dana laughed, but then the sudden reminder that Dr. Cavor was dead hurt sharply and the laughter died in her throat. "No. But I figured how to hack it temporarily while I got us low enough to breathe the pure O_2." She explained her nose-holding-self-pressurizing trick.

"Quick thinking! So are we safe now? Finally not going to die imminently?"

"No promises, but I think we're safer now than we have been since we accidentally launched this human fishbowl into space. Right now, I'm trying to get an idea of where we are. Thankfully, we're back below GPS satellites."

She turned on her phone, ignoring the voicemail alert that pinged, and activated the mapping function. She was delighted to find they were mostly in the right part of the world. Setting the Cavor School as her destination, she activated route guidance and turned the Beetle homeward, full speed. They were still high enough that the Beetle, despite its un-aerodynamic shape, could move along at a pretty decent clip in the thin atmosphere.

A little bit of calculation (and the navigation map) told her that it would be about 1:00 p.m. when they reached the Cavor School. Her phone told her the date — it was Friday, the day of the STEM contest judging. Unfortunately, the judging was set to start at noon. But Dana certainly wasn't a quitter. She would at least show up and make them tell her to her face how she had been disqualified.

From the notebook of Dana Dash

CHAPTER 31 — WELCOME BACK

They had spent days traversing the vasty deep between the Moon and Earth, but somehow it was the hours flying above terrestrial water and land that seemed to take an eternity. They spent much of that time discussing what they had seen and what they needed to do. It boiled down to: tell a teacher. Dana didn't like going to adults to solve her problems, but in the case of dangerous moon aliens and self-destructing-teacher-bombs, she thought this might be an excellent time to make an exception to that policy. But which teacher should they tell?

"I think we should go to Dr. Griffin first," said Dana. "She's in charge of keeping track of us at the dorm. We're going to have to explain where we've been anyway."

"She couldn't have been more uninterested in our whereabouts in the weeks before we left," said Noah. "I'd be surprised if she even noticed we're gone, unless Jayla asked about us."

"Well, who do you think we should tell?" asked Dana.

"I think Professor Moreau is second in command after Dr. Cavor. We should tell him first."

"Creepy-lizard-guy who has apparently been watching us for years? Dr. Cavor said those spy-lizards of his are practically illegal by Lunar Society law. Are you sure he's second in command?"

"I'm not sure of anything at this point, but that was the impression I got. He has the other private science lab. He's one of the Masters of a Discipline. Even the fact that he has a personal spy network and the other teachers tolerate it would seem to be evidence of his status."

"All right," said Dana. "Professor Moreau is Boss Number Two. We'll go to him first, unless we see Dr. Griffin first and she asks where we've been. Is that fair?"

"Sounds fair," said Noah.

And then, finally, they could see the school. They had made better time than they thought they would, which meant they were only going to be half an hour late for the contest judging. They had a good view far above Dash-on-Hudson, spotting the Cavor School grounds by the distinctive Zen garden on one end and the track on the other. Dana saw Myrtlegrove Manor, with its small green dome on the roof, and noticed that the lower round roof with the large skylight, Dr. Cavor's octagonal office, actually looked like a separate structure from above — as if the wings of the large house had been built around it, embracing it.

Dana flew the Beetle in lower and her ears popped from the increasing pressure. It was a relief to open the hatch and breathe in fresh, sweet, normally-breathable air. Noah dismantled Chip and Bosco's breathing setups as Dana steered the Beetle down, toward the main cluster of large buildings.

"I'm going to put us down right outside the gym, in the exact spot where we were supposed to have been camping all week, waiting for the judging today. Maybe we can convince them we were there the whole time and had our cloaking device on."

Noah laughed, "Or that our ship is really a TARDIS[97] and we did the week by skipping forward in time."

"Or that we switched place with alternate versions of ourselves from another dimension and we've now crossed back over into this reality."

Dana settled the Beetle down on the grass, outside the back door to the gymnasium where the STEM contest finals were being judged. She knew that they were probably too late, and that the winner — likely her least favorite finalist, Lemma Lei — was about to be announced inside the gym right now.

[97] The *TARDIS* (Time and Relative Dimensions in Space) is the time machine/spaceship/blue police box used in the really fun British science fiction series *Doctor Who*.

She thought they were going to have to get out of the Beetle and go inside to announce their return but, seemingly seconds after landing, the gym doors opened and five people walked out. As they approached the biosphere spaceship, Dana couldn't quite believe what she was seeing. The woman at the front of the approaching group locked eyes with Dana through one of the pentagonal polycarbonate windows and smiled as if they were sharing a joke.

It was Dr. Catherine Cavor — apparently alive and quite well.

Four of the five people coming to meet the recently arrived Beetle were the heads of each of the four disciplines at the Cavor School. The fifth was the gigantic woman Skadi, Master of Security, walking behind them and making the others look like small children being herded by a lone adult. Dana and Noah went to meet them, exiting through the airlock at the bottom of the biosphere.

"How can Dr. Cavor still be alive?" asked Noah in a stage whisper. "And how could she have possibly gotten back from the Moon? Ahead of us, no less!"

"I don't know," hissed Dana. "Play it cool. Don't say anything until we know more about what's going on here."

They slid out from under the Beetle. Noah was so happy to be back on Earth that he lingered on his hands and knees for a moment and gently kissed the ground. Dana, however, was on her feet immediately, looking up at the daytime sky, thinking about the stars behind that blue screen, and wondering if she would ever get to see them up close again.

She took a deep breath, and was startled by the smell of clean, fresh air. Her nose had become deadened to the smells in the Beetle: stale air, unwashed bodies, spilled waste water, soft green cheese fungus, and the burnt firecracker smell of moondust. It had all faded into the background after a while. But the sudden lack of that horrible mixture of smells was incredibly noticeable, stopping her in her tracks. While she was processing the fresh summer breeze, the five staff members arrived in front of the two children.

"Dana, Noah," said Dr. Cavor. "Sorry to keep you waiting. We got started a little late, but I made sure we came to look at your project first, as I know that, after a week of confinement, you would be eager for the taste of freedom. The other three heads of the STEM disciplines are here as judges. Master Skadi, whom I am sure you have met, is filling in for me as the fourth judge. Since I am Head of School, I am recusing myself from judging the contest, unless there is a tie that needs breaking."

Dana noticed that each judge was wearing their discipline color and that Master Skadi, like Dr. Cavor, was dressed in all blue.

"Now, tell us. How did your project work out?" Was Dr. Cavor pretending that Dana and Noah hadn't gone anywhere — that they had been here testing the biosphere the whole time, as planned? That meant they were still in the contest! Of course, Dana was happy to go along with it, but she couldn't help but wonder what the motivation was. It also left the big question of how Dr. Cavor was even here, back on Earth, still alive and breathing. But all that could wait. Time to try and win this contest.

"Well, it certainly wasn't as easy as I thought it was going to be," said Dana in the understatement of the millennium.

"Yeah, we had some problems with the air recycling system," said Noah. "Totally my fault."

Professor Moreau looked surprised. "The cyanobacteria did not perform as expected?" he asked. "It looked very promising, as described in your project proposal."

Professor Uachen and Dr. Taxidiótis stood behind Professor Moreau and Dr. Cavor. They both leaned forward, also interested in the answer.

"The cyan-aid — I mean cyanobacteria — worked perfectly," said Noah. "Better than expected, really. But I tried something else that didn't work out well. I used cryptophytes that could both digest organic material and do photosynthesis in two of the tanks. I was experimenting with breaking down solid waste to get us closer to a truly self-contained biosphere. I also did two tanks of mixed crypto and cyano. So two-thirds of my air farm had the crypto in it and... well, something went wrong there.

"When those tanks all went bad, we basically ended up with only one-third of the CO_2 to O_2 conversion capacity we should have had."

"So you had to open the door and let in air?" Dr. Taxidiótis asked.

"Oh, no," said Noah. "Dana figured something out. We had solar cells and some electrical stuff. She generated oxygen from our water tanks using electrolysis."

"A fine example of Technologist resourcefulness," said Skadi, winking at Dana with her non-eyepatched eye. At least Dana thought it was a wink. Maybe a long blink? How could she tell?

"So you went a whole week without taking in additional air?" asked Professor Uachen, the mathematician.

"Well, umm.... "Dana was thinking about the fact that the Beetle had been opened on the Moon inside the pressurized moon bug tunnels and they had breathed plenty of lunar air.

"Yes. Tell us." Dr. Cavor looked right at Dana, her iridescent eyes once again hidden behind her glasses. "Did your biosphere remain separate from *Earth's* atmosphere for the entire week?"

"Oh, right," stammered Dana. "Um, correct. We didn't take on any new air from *Earth's* atmosphere for the whole time."

"Very good." said Dr. Cavor. "And what is the main lesson you will take away from your project?"

Dana thought for minute — it was a good question, but not the sort she had expected. "I guess — I guess I learned that dealing with the basic needs of human bodies is frustrating and sometimes really gross and turns out is a much bigger part of space travel than I imagined."

Dr. Cavor smiled. "Yes. The human body is indeed annoying and often makes it difficult to accomplish anything truly interesting."

Dr. Taxidiótis had walked over to the Beetle and was peering through one of the lower windows. "What is the Yagi antenna for?" she asked.

"Oh, well, we had some time to kill this week, and I had a couple of ham radio handsets to play with," Dana said. "I decided to try to see how far away I could talk to someone."

"Ooooh! I love ham radio. How far away did you manage?"

Dana was about to tell her about talking to Conrad Cartier in London, but then was worried that if the professor knew radio stuff, she would have to explain how she managed to orient an obviously fixed-in-place antenna for an EME bounce. She was saved by Bosco, whom they had forgotten all about when they saw the contest judges coming. He jumped down from the open airlock hatch and wandered out onto the grass, sniffing the ground.

Professor Moreau said, "Oh, you had your dog in there with you? Was that in your original air production plan?"

Bosco looked up with a startled expression when he heard Professor Moreau's voice, gave out a squeal-bark, bolted back under the Beetle and jumped back up into the hatch.

"That was strange," said Dana, looking from the Beetle and back to Professor Moreau. "He likes everyone, usually."

Professor Moreau had also been surprised by Bosco's reaction. He stroked his wispy beard thoughtfully. "How long, exactly, have you had that dog?" he asked.

Dr. Cavor interrupted. "We must look at the other projects now. We are running late already and we have to announce winners back at the manor house at four p.m." She began to walk away.

Professor Moreau looked oddly towards the Beetle where Bosco had taken refuge and held out his hand. A small flying lizard flew from the Beetle and landed on his outstretched palm. Had it come from inside the Beetle? Could they have had a stowaway spying on them this whole time? Dana couldn't be sure. Then, perhaps somewhat reluctantly, Professor Moreau turned and followed the other judges who had immediately taken Dr. Cavor's cue to head back to the gymnasium.

"Did that lizard come out of the Beetle?" Dana asked her friend. "If so, color me annoyed."

"Me too," said Noah. "I'd hate to think we had a captive lizard with us that whole time and I didn't get a chance to document it at all. I really want to know the science behind what he does with those things. Did you see it come to him, like he controlled it with his thoughts?"

Dana was more upset with the invasion of privacy than the missed research opportunity, but she said nothing.

Before following them, Dana and Noah retrieved their lab books and checked to make sure Chip was all right in his cage, and then Dana got Bosco's leash and tied him to the landing gear leg. As they walked to the gymnasium, Dana realized that they were both unconsciously walking in the same gliding hop-step they had done on the moon, but in the much lighter gravity there. How was that possibly still working on Earth? She took a little extra hop on the next step and floated easily an extra meter through the air.

"Noah, how is it that we're still in low gravity?"

"Huh?" Noah stopped and turned to her, confused.

"Why can I still do this?" she asked, and jumped straight up. At the top of her long, slow motion leap, her extended feet were several meters off the ground. She fell slowly back to the ground. "Are we really back home? Could the moon bugs have tricked us with some sort of alien technology? What if our return to land on Earth was all faked in some sort of sound stage on the Moon!?!"

"What!? No way! That's crazy." Noah was obviously baffled. But not for long — a moment later he exclaimed, "The water!"

"What water?"

"We neg-gravved all the tanks. The bacteria tanks and the drinking water tanks. All of it. The water we've been drinking all week. It was all converted to neg-grav." He jumped a similar distance into the air himself, laughing. "This is Earth. But we've been drinking neg-grav water long enough to have replaced a large percentage of our bodily fluids." He stopped, looking worried.

"What is it?" asked Dana. "Is something wrong? Are we ok?"

"No. No. We're fine," said Noah. "I was thinking that the human body is well over fifty percent water by mass. A few more days of drinking converted water and the balance might have tipped the other way. We didn't even think about it but we could have found ourselves floating off into space right now, rather than able to jump really high."

Dana looked up at the sky and shuddered, imagining what it would be like to keep falling up and up... forever.

CHAPTER 32 — JUDGMENT DAY

Dana shook herself. "Ok, let's not to think about what other bad things could have happened but didn't. Right now we need to act like we're still in this contest and look at our competition's finished projects." She smiled and said, "But if we have time later, we can try out for a pro basketball team." She continued her skip-glide towards the gym and Noah followed her.

When they entered the gymnasium, the judges were already talking to the math kids about their cryptography project. The giant Master Skadi was bending far down to examine information on their computer screen, while Professor Uachen was engaged in animated conversation with the twins, most of which was incomprehensible to Dana. She was used to Toma and Tevi constantly finishing each other's sentences, but the professor was also doing it, so much so that it wasn't even clear who was asking the questions of whom. The words were such a dense string of mathematical terms that Dana might have been lost even if the conversation had been one single speaker. She decided, not for the first time, that she really would learn more math. Math was at the root of everything interesting.

In due course they wrapped it up and the judges moved on to talk to Jayla, while Noah and Dana came over to talk to the math kids.

"We missed some of your talk," said Dana. "How did it go? Did you get your encryption system working?"

"It works, but it didn't deliver the expected..." said Tevi, "...speed increase over an elliptic curve algorithm," said Toma.

"How did the flight plan work out?" asked Toma. "Did we get it right?" asked Tevi.

"Well, we're here," said Dana.

"We did come in a little fast," said Noah. "It got pretty hot."

"I wonder what went wrong," said Tevi. "I thought we had it," said Toma.

"I'm pretty sure I know what the problem was now," said Dana. She jumped off the floor, although not as high as she had earlier, so as not to draw everyone's attention. She only went slowly up and down about a meter but it was enough to widen the eyes of the two mathematicians. "We were drinking neg-grav water the entire trip. Transferring power out of the brakes," she explained.

"Ah," said Toma. "Of course," said Tevi. "So it's real," said Tevi. "We didn't really believe you," said Toma.

At Jayla's table, her robotic arm came to life with a loud warning beep. They all turned to watch. She was wearing a golden headband, the same color as the arm, with electronic components on it. The arm reached out and picked an egg out of a carton, holding it up for the judges. Then the arm swiveled and reached out to a bowl, cracked the egg sharply on the edge and deposited the contents into the bowl, single-handed, like a master chef. Then it put the shell back into the carton and did the whole thing again. Finally it gave the judge a thumbs-up, and everyone clapped.

The judges moved on to the science kids, while Dana, Noah, Toma, and Tevi all went over to congratulate Jayla.

"That was amazing!" said Noah.

"Definitely impressive," said Dana.

"It was actually a bit of a cheat," said Jayla. "Picking up the egg was controlled from here," she tapped the headband, "but removing the yolk was a set of preprogrammed moves — I initiated it. Dr. Taxidiótis knew, but said it would be flashy and impressive anyway."

"Very impressive," said Dana. "I hope the science kids don't have anything that cool. If you win, I'll still be totally jealous, but I don't think I'll even be able to stand it if Lemma wins."

Jayla nodded, then sniffed the air. "Phewww. You two smell like you've been trapped in a hamster ball with two other animals for a week." She grinned, and Dana shrugged. First chance she got she was taking a shower.

They all wandered over to look at Team Science's offering. It was, unfortunately, quite impressive-looking. They had surrounded their whole display area with a fine mesh wire cage. Their large table was covered with a shiny gray table cloth that went down to the floor, which was covered with the same material all the way to the edges of the cage. On the table was a bewildering array of interesting equipment and some animal cages containing white rats. Prima was standing inside the cage, behind the equipment table, while Lemma was in front explaining the setup to the judges.

"We had to surround our equipment with a Faraday cage to isolate our experiment from outside radio frequencies. Even the floor is covered with a conductive fabric to make sure we have total radio frequency isolation. Initially, we were getting so much interference at our lab that it seemed almost like there was active jamming in the specific frequency range that we discovered is produced by vertebrate bioelectric activity.

"But once we were working in an isolated environment, we were able to reproduce data consistent with the hypothesis that rudimentary radio communication exists in all vertebrates. It has long been known that many animals respond to electromagnetic fields. For example, some birds use a sort of built-in compass for navigation during migration. It is also possible that other species, like fish and turtles that return to the same places they were born to spawn, do something similar. However we've been able to demonstrate conclusively that most vertebrates are capable of sharing information over radio frequencies generated by their bioelectric fields."

Dana whispered loud enough for the other kids around her to hear, "Tell me that she didn't say they discovered ESP..."

But Lemma continued before anyone responded. "Despite how useful such information sharing would seem to be from an evolutionary standpoint, the broadcast capabilities in these frequencies, while decidedly present, are so weak as to be vestigial. But when we boosted these frequencies, the receiving capabilities were still there, generating clear neural impulses up the spine from a site between the fourth and fifth lumbar vertebrae. That is the location of a cluster of nerves that may be what many paleontologists once thought was a second brain in dinosaurs."

Dana was doing her best to make sense of what Lemma was saying, "The spinal cord acts as an antenna?" she whispered to herself.

Lemma said, "Here we have two identical mazes." Prima uncovered the maze setup on one end of the table, gesturing like a game show model, and Lemma continued, "First we release a rat into the maze that has never seen this setup before." Everyone watched as the rat moved around the maze slowly and randomly. "Then, in the other maze, we have a rat who knows this configuration well and can run it fast for a reward. You see the little helmet it's wearing? That detects the natural radio frequencies it's generating and this equipment here amplifies those signals by several orders of magnitude. The other rat is unaugmented. It will naturally receive and respond to these signals."

Lemma retrieved the uneducated rat and returned it to the beginning so they could start both rats at the same time. "Now watch what happens."

Both rodents now moved in almost perfect synchronization through the maze to get the cheese at the end, and Dana knew she had been beaten. If what she saw here was real, they hadn't witnessed a winning science fair project, but a truly revolutionary, probably world-changing, discovery.

When Lemma's demonstration was over, she allowed Prima to join her outside the cage to answer questions. Dana was too upset about how good the project had been to listen to what was said but, as the judges started to leave, she tried to put it out of her mind and see if she could get Dr. Cavor to tell her what had really happened on the Moon. She caught up with the Head of School and asked, "May I have a moment to talk to you in private? I have a few questions."

Dr. Cavor replied, "Not right now, Dana. No time. I promise to talk to you at length and answer any questions you might have. But you will have to wait until after the STEM contest winners are announced. I will see all of you in a half hour at the manor house. The judges will have made their decisions by then." Then Dr. Cavor followed the judges out.

"Well," said Noah, "it's good that we will be getting some answers eventually and a half hour is enough time for a quick shower. Which we desperately need. Come on, let's go."

Lemma must have overheard Noah's comment from where the other kids were congratulating her and Prima and called out, "Good idea! We can smell you both from here."

Dana, her face burning, followed Noah out the door.

They stopped back at their dorm rooms for the showers. Dana finished hers first, too excited to really enjoy the hot water, despite having missed it for a week. She dressed quickly and went to find Noah. When she knocked on his door, he yelled for her to come in.

"It's unlocked," he called out from the bathroom.

"Are you almost ready?" she said as she came inside, speaking loud enough to be heard through the bathroom door.

"Almost. I'm done with my shower. Give me a minute." She heard him clunking around in the bathroom a bit.

Dana started to think about what she wanted to ask Dr. Cavor, and then what she needed to tell her. She had seen those scarecrow-hat-controlled deer on campus right before they launched and had never told anyone. That picture of a scarecrow on the Moon had been similar, and pretty ominous. That thing might be a serious threat and the adults should know it was here.

Her train of thought was interrupted when Noah suddenly yelled, "Aaak! What the...? Oh no!"

"What happened?" she called out to him. "Are you all right?"

"Not hurt." He called back. "A little problem here. Um... it's a bit embarrassing."

"Oh?"

"Um... It turns out that going to the bathroom has its difficulties when one needs to eliminate neg-grav bodily fluids. I'm afraid I'm going to have to take another quick shower."

Dana tried, not very successfully, not to laugh. "Thanks for the warning. I'll be sure to go to the Beetle and grab the suction tool we improvised in almost-zero gravity before I try using the toilet."

While she waited for him to shower off again, Dana checked the voicemail on her phone.

"Hey Danasaur!" her mother's voice said. "I'm running a little late. I'm not going to make it to your final ceremony. I'm so sorry. We had some problems on vacation. Your dad and Noah's parents are going to be on an even later flight than the one I managed to get. We're all so proud of you and want to be there, but we called the school and it's ok for you and Noah to stay late tonight until I can pick you both up. I'll be there soon! And I'm crossing all my fingers and toes for you!" Dana's mom was obviously trying to sound upbeat, but was trying a little too hard and it was clear that whatever vacation snafu[98] had occurred, it had her very stressed.

When Noah was ready, she told him about the messages and he checked his phone to find a similar message from his mom, saying he should get a ride home with Dana. They agreed to put any concerns out of their minds for right now and they headed over to Myrtlegrove Manor.

"Don't you want to race over there?" asked Noah.

"No thanks," said Dana. "The only thing I'm hoping to win is this contest. And I want to win it with you. I'm not feeling like I need to beat you at anything right now."

"What makes you think you can?" asked Noah with a smile.

"Oh, I think I got in more practice doing low-grav running than you. But I don't think it's a good idea for anyone to see the way we're moving right now. Let's take it slow and try for the most normal, least silly walking possible."

---

[98] *SNAFU* stands for "Situation Normal: All Fouled Up." It means that yep, things just went really wrong, again. It was originally a World War II military acronym meaning "Status Nominal: All *^&^% Up." War is really a terrible thing so that's why soldiers swear like that.

When they reached the manor house, there was a similar setup to the first day of the contest. Rows of chairs with a central aisle faced the large staircase with the long curved railings at the back of the large entry hall. There was a podium with a microphone set up at the foot of the stairs. The other contestants were already standing up there, so Dana and Noah walked up the aisle to join them, doing their best to pretend to be walking normally despite their reduced weight. They found a place to stand next to Jayla, who was still wearing the control headband from her project.

"Oh look, its Moon Girl," said Lemma to Prima, just loud enough for Dana to hear, "She is about to wish she had stayed in space."

Dana was surprised to see that there was a larger crowd than when the finalists had been announced a month ago. Mixed among the very ordinary looking people from what Dana assumed was the town, Dana noticed that the room filled with an unusual amount of bright colors, loosely sitting in the same sections mixed in with the much more drab clothing of the townspeople. The color groups were red, green, blue, and black — the same as the colors of the STEM disciplines. It had the effect of making the audience look like sports fans on game day, dressed to show support for their favorite team.

Lambda Lei was sitting with her mother, both dressed in dark red. Professor Moreau was also in the red section, next to the Leis. Dana scanned the cluster of green, wondering if Ms. Astrulabi would be there. *Wait till she hears how I learned to use the astrolabe!* she thought, before wondering if it was wise to tell anyone at all what had happened. However, Ms. Astrulabi's trademark green headscarf was nowhere in the gathering crowd that Dana could see. She did see Professor Hetzler sitting next to Dr. Griffin, both dressed not in colors, but in white. Against the bright hues of the other audience members, they almost seemed to glow.

The skulky guy in the mismatched suit and bowler hat was there again, slinking around the side of the room. Dana watched as he sidled up to the blue quadrant, pencil and notebook in hand. Skadi, who had been sitting in the front with the faculty, must have seen him too, because she rose from her chair so abruptly the chair tipped over. She walked towards the skulky man, reaching him in a few large strides. The crowd was still talking, but Skadi's voice carried above the din.

"I have heard tell of your wagers on these younglings, Frumver Frigjar. They are not horses for your sporting pleasure. Begone with you."

The man chuckled. "Oh, Skadi. You never were much one for fun." Seeing Skadi standing firm before him, her large arms crossed in front of her, he bowed low and tipped his hat. "As you wish, my dear. I was planning to depart anyway." Before Dana could even blink, he had slipped smoothly out the door.

As the mysterious Lunar Society bookie disappeared through the doorway of Myrtlegrove Manor, other apparent fans were still arriving. A bulbous shadow appeared — it was Principal Peters, followed by Matt McCaws and the man she knew was Matt's father. She'd have recognized Mayor McCaws anywhere, because his face was plastered on nearly every yard and storefront in town. Matt was so completely absorbed with his phone, oblivious to the crowd, and probably annoyed at being dragged along with his father and uncle, that he didn't even look up as his father guided him to a seat near the back.

Dr. Cavor had been off in the corner, talking with all of the judges except Professor Moreau. As Skadi returned to the front of the room to stand with the judges, Dr. Cavor came over to the microphone and tapped it to get everyone's attention.

"Welcome welcome. Thank you so very much for attending this year's Cavor School STEM contest finals," she said. "The members of the winning team will each receive a full scholarship to the Cavor School and have their names engraved on the official plaque with all the winners since the contest began. Members of the other teams are also, at the discretion of the judges, offered admission to the school and every attempt will be made to provide them with other existing scholarships and need-based financial aid.

"This year, we had four outstanding proposals by four very dedicated teams whose execution of their projects far exceeded our expectations.

Therefore, I am delighted to announce the judges have agreed that all four teams will be invited to attend the Cavor School next year."

The audience broke into enthusiastic applause. Dana felt nothing for a moment, but as the applause continued, she felt a warm flood of relief. She would likely be able to study advanced science. Dana was particularly happy to hear about the possibility of other financial aid. She had the feeling that Lemma and Prima both came from families with money, but even if they won what they didn't need, there was a chance that she, Noah, and the rest of the contestants would still be able to attend next year.

Down the row of finalists, everyone grinned happily. As each of them met her eye, they smiled back at her. Even Lemma seemed to share a moment of human contact with her. It almost seemed like it didn't matter who won the contest now.

Dr. Cavor wasn't finished. "However, the contest can have only one winning team."

Dana tensed. Noah reached for Dana's hand and held it tightly. Everything they had just been through had been in the service of this moment. She wasn't exactly sure how she felt, or how she was supposed to feel. After moon bugs and a spaceship ride and re-entry into orbit and Dr. Cavor still alive — how much did the contest even matter anymore? She wasn't sure how she would react to the next thing Dr. Cavor would say.

She was almost sure her and Noah's project hadn't been as good as Lemma and Prima's. She wasn't even sure if it was better than Jayla's — that robot arm had been wicked cool. And she didn't really even understand the math kids' projects, which in itself was a bad sign.

Dr. Cavor continued, "Final voting is, of course, done by the electoral members."

It is? The judges only decide the admissions? Who are these "electoral members" who can vote?

"I trust that everyone has had time to look at the presentations made by each judge and cast your votes if you are a member, so I will now check the tally." Dr. Cavor pulled a tablet screen from her pocket.

After all the old-fashionedness of this high-tech school, Dana was surprised to see actual tech being used.

"...oh, this is quite unusual," said Dr. Cavor. "We have a tie. I have only seen that happen once before, and as Head of School, it is my responsibility to break it once again."

Dr. Cavor got to decide? She was Master of Technology! Dana and Noah's advisor. She would vote for them, for sure! They were actually going to win this! Oh, assuming Delta was one of the teams that tied. They might not be... her heart sank a little, but, still hopeful, she continued to listen.

"And this is not a very hard tie to break. In my judgment, the team who clearly deserves to win this year's STEM contest is — "

BOOM!

She was cut off abruptly as the large entry doors of Myrtlegrove Manor crashed open, echoing loudly in the hall, despite the number of sound-absorbing bodies crowded into the limited seating. There in the doorway stood a tall figure covered in torn clothes, rags, and plastic bags. The back part of a deer body protruded from the man-shaped trash collection in front, and the antlers of the buck protruded out of the man-figure's shoulders like jagged, bony wings. It regarded the crowd through the wrappings around its misshapen attempt at a face. On its head sat the tattered remains of the garden scarecrow's floppy hat. From beneath that hat, the Scarecrow's single glowing green eye silently judged them all.

CHAPTER 33 — BOSS FIGHT

Everyone in the entry hall turned and looked at the door as the Scarecrow-centaur made its entrance. As the Scarecrow fully entered the hall, Dana could see that it had a connected train of deer in tow. *The deeripede!* The Scarecrow moved like a snake, the deeripede trailing behind like a game of whiplash. It moved towards the back row of chairs, and the cavernous manor hall filled with sounds of confusion. People scrambled to move out of the way, knocking over chairs and stumbling into each other and falling. The whipping-deer-tail blocked the exit. The Scarecrow reached out its arms and touched two people, one on either side of what was left of the aisle of chairs. One woman screamed — but the scream died in her throat as the Scarecrow's arm touched her shoulder, and her face was instantly drained of all expression. The woman had become part of the deeripede. She reached her arms out and touched two other people, and those people did the same, and so on. People were panicking more loudly now — screams and cries of fright — but the noise was decreasing. Each row went dead silent when they were touched by the people behind them.

The exit was now thoroughly blocked by the outstretched hands of the horde and the whiplashing tail of the deeripede. The children began to back up the stairs, away from the encroaching zombie outbreak. Only Jayla, in her wheelchair, had nowhere to retreat. And once upstairs, the others would be just as trapped.

Master Skadi picked up a large rectangular table that had been pushed up against the wall and moved to get between Dr. Cavor and the crowd. Dr. Cavor called out, "Protect the children!" The other teachers snapped out of their initial shock. Dr. Taxidiótis pulled a slide rule and a drafting compass out of her pocket protector and advanced toward the crowd. The items in her hands began to glow green. Professor Uachen cupped his right fist in his left hand and pushed them outward. Both hands began to glow with a black light that seemed to come through the skin as he raised them overhead.

Master Skadi reached the front of the seating area, between the crowd and where Dr. Cavor stood in front of the children by the stairs. She raised the table as a shield between them. The front edge of the crowd had not been turned yet, and Dana could see the look of panic in many eyes, including Lambda Lei and her mother, as they realized that they were being pragmatically sacrificed by the giant woman. Professor Moreau was also trapped there. Dana watched his face go from angry confusion to a complete blank as a hand reached out from behind to touch him.

As soon as Master Skadi was between the children and the crowd, Dr. Cavor acted. The middle-aged academic woman let out a loud "Kiai-Ya!" as she leapt into the air, arcing above the aisle between the two now silent halves of the crowd. At the top of her flight arc, her raised fist began glowing with a blue light, and it came down with her full body weight behind it, towards the head of the Scarecrow.

At the last second, the Scarecrow seemed to avoid the blow by splitting in half. Its right arm and shoulder fell apart into rags and small writhing things. All the people on that side of the aisle also collapsed, their puppet cords cut. The larger part of the Scarecrow collapsed the other way, falling into the crowd there, which maintained its eerie, silent cohesion. Dr. Cavor's fist came down on the revealed buck's head, right between its antlers. With a bright flash of blue light, the deer crashed to the floor unconscious.

Arms reached out of the still upright part of the crowd to touch Dr. Cavor. Dana expected to see her also fall under the Scarecrow's spell, but whatever it was that the Scarecrow did to take control, it didn't work on Dr. Catherine Cavor. She angrily shoved people aside, reaching into the crowd after the mass of rags that had been the Scarecrow's head, but it was eluding her in the throng of people and folding chairs.

Then both of Dr. Cavor's hands began to glow blue, and each person she touched fell to the floor.

The drafting compass in Dr. Taxidiótis's left hand had produced a large round glowing green disk, which she used as a shield to keep grasping arms from reaching her as she pushed the zombies back. The slide rule in her right hand was being used like a sword, darting out to touch zombie after zombie. Each person touched collapsed to the ground with a shower of green sparks. Professor Uachen was standing to her left, also gaining partial coverage from Dr. Taxidiótis's shield to the right and Master Skadi's table to the left. His weirdly black glowing hands darted out with lighting speed to tap foreheads and temples, apparently too fast for the contact to cause him to be taken, and with each strike, another zombie fell unconscious.

Dana was watching the teachers fight, fascinated, when she felt a tug on her pant leg. She looked down and saw Chip, looking at her with an expression that said that he would have something to say if he could talk. She pulled the stone from her pocket and popped it into her mouth.

"The Scarecrow is after us and the stones. Up the stairs. We can get out the open back window I came through," Chip said.

"You call it the Scarecrow too?" Dana mumbled around a mouthful of stone.

"You are mostly talking to yourself, remember? Now move!"

But the four school staff members were winning this thing. They had all demonstrated amazing, unexpected abilities, and were using them to effectively contain and reduce the zombie threat. It looked like the thing scrambling among legs and between bodies — the core of the Scarecrow — would soon have no more army of zombie soldiers to hide among.

Then everything changed. Dana saw Zombie Moreau pinned by Master Skadi's table. He threw back his head and opened his mouth wide in a silent scream. Instead of sound, four little dragon lizards shot out of his mouth.

Isn't anyone else seeing this? Dana looked around, but the other students seemed frozen on the stairs. She tried to yell a warning as each of the flying lizards followed an upward and outward path that curved back down behind each of the four effective combatants. But the stone slipped into the back of her throat and the warning came out as a gagging cough. Each of the little lizards found its target at the back of a teacher's neck. All of their faces, except for Dr. Cavor's, went blank. The lizard that landed on Dr. Cavor had no noticeable effect on her.

Dana realized in growing horror that Zombie Moreau's link with the lizards was being used to transmit the contagious mind control. The Scarecrow was no longer limited to needing an unbroken chain of contact to control its victims. And now three of their previous defenders had been converted to attackers.

She swallowed the strange stone once again and managed to yell, "It's Professor Moreau! His lizards!"

Dr. Cavor understood at once and leaped high into the air again, her blue glowing fist raised high, targeting Zombie Moreau. But as she came down, the now-controlled Zombie Skadi swung the long table like a baseball bat and hit a home run. Dr. Cavor flew backwards, across the entire entry hall, and out through the open doors. Zombie Skadi dropped the table. She turned a blank one-eyed gaze towards Dana and the rest of the children, and began lumbering towards the foot of the stairs — *straight towards Jayla.*

"Am I just talking to myself here?" asked Chip. ***"I told you to go. Now go!"***

Dana moved, but not away from the threat. She jumped back down the stairs in a slow-motion low-gravity glide. She landed right behind Jayla's wheelchair as the giant Zombie Skadi reached out for Jayla with both hands. Dana expected her to be lost to the Scarecrow then and there, but instead, Jayla reached up to touch her headband, and it lit up. Long, multi-jointed robotic arms unfolded from the thick back of the wheelchair. Jayla's metal super arms thrust forward in front of her and grabbed the giant woman's wrists in a powerful robot grip.

Zombie Skadi jerked her arms back, trying to break free of the wheelchair that had suddenly sprouted arms and become a formidable exoskeleton. Jayla and her chair were lifted off of the ground. Dana grabbed at the back of the chair to keep it down, but instead Dana was lifted into the air too.

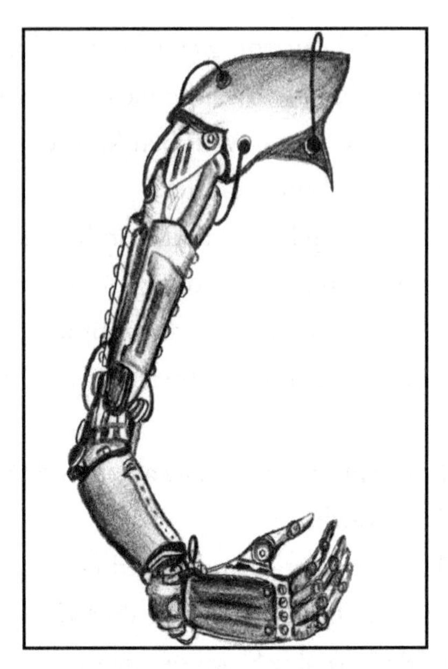

From the notebook of Jayla Jones

Jayla's finger was still on her headband, pressing it tightly as the chair was jerked side to side by Zombie Skadi trying to free herself from the robot arms. Then Jayla made a motion and the chair arms pulled backwards, holding the giant in place. A sort of robotic foot shot out of the bottom of the chair and kicked the giant woman square in the chin with uppercut worthy of a prize fight. Zombie Skadi flew backwards, pulling the robot arms out of the back of the chair. She slammed into the rows of chairs and went to the ground hard.

Jayla, her chair, and Dana fell to the floor. Somewhat slowed by Dana's partial neg-grav state, they managed to come down on their feet and wheels, and Dana proceeded to pull Jayla backwards up the stairs.

"What are you doing? I got her. And my grandma is in that crowd!" yelled Jayla.

"Maybe so," said Dana, "but what about the rest of them? We need to fall back and rethink this." The tangled crowd among the folding chairs had reorganized and was heading towards them. They had also reached across the aisle, and the previously downed zombies on that side were getting up again, reconnected to the mass that contained the Scarecrow. Dana could see Zombie Moreau, connected to Lemma's mother and sister, and noticed that the bundle of filth that was the Scarecrow had climbed onto his back.

The Scarecrow had found itself a new favorite mount.

Zombie Taxidiótis and Zombie Uachen had also both taken a step in the direction of the stairs when Dr. Cavor came back through the door, both hands glowing bright blue. *Now she's going to show them,* thought Dana. Then she noticed that Dr. Cavor's head was lolling to one side, her neck misshapen. Was she still standing upright and ready to fight despite what looked like a broken neck?

Before Dr. Cavor could move, Zombie Taxidiótis raised her slide rule and pointed it towards the doorway. A blinding green beam of light lashed out and burned Dr. Cavor cleanly down the middle. Sparks flew, burnt wiring was exposed, and gears and electronic parts went flying everywhere. Dr. Cavor fell to the floor in two pieces — stone dead. Again.

So, not an alien; she was a robot! Of course, it all made sense now. Dr. Cavor hadn't needed to make it back to Earth. She had left a backup copy in her place in case she couldn't return.

"All right now, how many times am I going to have to tell you to run?" Chip asked, as Zombie Taxidiótis turned towards the children, the incredibly deadly slide rule still glowing brightly in her hand.

Dana increased her pace, and the "leg" under Jayla's chair began to demonstrate what was the more likely purpose for which had been designed. It extended to push as the "foot" part rotated in a motion that lifted the chair backwards and up, one stair at a time. With Dana pulling and the foot pushing, Jayla's wheelchair managed to ascend the stairway at a very decent speed.

Dana and Jayla's movement up the stairs, past the other children, broke the amazed trance they had been in while watching the unbelievable fight. They came close behind, following quickly up the stairs, down a hallway, and into a small office with a large old-fashioned floor-to-ceiling window, which was wide open. Chip wasted no time, leaping over the desk and through the window, sailing lightly to the ground below.

Dana called out, "Toma and Tevi, lock the door and block it with the desk. Prima and Lemma, pile things on the desk. That chair. The file cabinet. Noah, help me with Jayla at the window."

Lemma said, "Who made you boss? Why should we — "

"Shut up!" interrupted Prima, who was already moving the file cabinet. "Do what she says or don't. But please just shut up!"

Lemma looked furious, but she shut up and helped Prima tip the file cabinet onto the desk.

Dana moved onto the windowsill and directed Noah to do the same, positioning Jayla in her chair between them on the large window sill. "Ok, jump!" she ordered.

They held Jayla's chair together from either side as they jumped, and their lower gravitational mass worked to also slow Jayla's fall.

Lemma was at the window now, yelling, "See! They are leaving us behind. I knew we couldn't trust them."

Dana motioned to Noah to lay Jayla on the grass, then pointed up to the window and said, "Again. Jump!"

The two of them jumped together, sailing up to catch hold of the windowsill. In unison they reached up and grabbed Lemma, who was facing back towards the other kids in the room.

"Why would you follow her orders. Now we're going... Hey! No! What are you doing? Stop!" she yelled as Dana and Noah neatly defenestrated[99] her. All three children fell-floated to the ground, where they deposited her somewhat gently. Although, perhaps not as gently as they could have.

They repeated the process twice more, smoothly, for each of the math kids. But on the final trip, Prima was still piling random objects on the

---

[99] *Defenestrate* means "to push or throw someone or something out a window."

desk — what looked like a shredding machine was going on top of the roller chair as they arrived at the window. "Come on, Prima," called Noah.

As she turned to come to them, there was a crash. Zombie Skadi's huge right fist, with its Euler's Identity math tattoo, punched through the door as if it were styrofoam. The filing cabinet pitched off the desk and struck Prima's leg. There was a pop, and Prima screamed. She fell against the windowsill. Dana and Noah hauled her through the window and took her to the ground as gently as they could.

Dana saw something moving quickly through the grass towards Lemma. Leaving Noah supporting Prima, she stepped on it — not hard enough to kill the little lizard, but to pin it with her shoe.

"Watch out for these little lizards," she said. "If one touches your skin, your mind will be gone. You'll become a zombie-you. Noah, do you have one of those little plastic boxes?"

A good biologist even in a crisis, Noah produced a small specimen bag from his pocket. "Will this work?"

Dana nodded and cupped the wingless lizard with the edge of her t-shirt so as not to touch it directly, and deposited it inside the bag. "There you go, Noah. Now you have Mystery Lizard Five to study."

"Cool. This one is shaped like a skink," he said.

"Fascinating, I'm sure, but you should really head for the Faraday cage in the gym. It will block the signal to the lizards," said Chip.

"Good idea, Chip," said Dana. "Run for the gym! Prima and Lemma's project has a Faraday cage. It will block the signal to Professor Moreau's lizards. Come on!"

"I don't think I can even walk," said Prima.

"Put her on my lap!' yelled Jayla.

Noah helped Prima onto Jayla's lap and almost everyone started towards the back of the school building and the gymnasium.

"Wait!" yelled Lemma. "I think we should head for the dorms and pull one of the fire alarms."

"I don't think the police or firefighters are going to be able to help with this," Dana yelled back and kept moving.

"I don't know what's going on here, but..." said Toma. "...it sounds like maybe Dana does," said Tevi. "Maybe when things calm down..." said Tevi. "...she will even explain why she seems to be getting her ideas from the fox," said Toma. The twins ran towards the gym after Noah and Dana.

Crashing noises could be heard from the room above as Zombie Skadi broke through the door and began clearing the piled-up furniture from her path.

"Heyyyywwwait for me," called Lemma as she hurried after them.

CHAPTER 34 — CAGED ANIMALS

The children, five running, one driving, and one passenger, made their way quickly across the campus to the gym. The fox had gotten there first, but with four legs, that was really no contest. Dana got there next. She stopped and held the door for the others.

When Noah got there last, arriving behind Jayla and Prima, he said, "What? No Dana Dash trash talk victory speech?"

Dana laughed and said, "You know what, it didn't even occur to me. Maybe with everything that's gone on, I've done a little bit of growing up."

"Great," said Chip, ***"You're all grown up. Now, listen to your imaginary friend and go hide from the boogeyman under the table."***

Prima made a pitiful noise, perhaps to remind them that she had been injured and would appreciate an end to the uncomfortable ride at some point, while Jayla continued to hold the other girl on her lap with no complaints.

"Come on," said Dana, "Let's get ourselves hidden while we figure things out."

The entire collection of STEM contest finalists fit under the science kids' project presentation table, although Jayla had to climb out of her wheelchair to do so. They huddled there, protected from the outside world by gray fabric hiding them from view and a wire mesh cage which would hopefully prevent the Scarecrow from controlling anyone through the little lizards. A few minutes after they got themselves situated, they heard the gym door open again and they all remained as silent as possible. Very heavy footsteps that had to be Zombie Skadi's echoed through the gym. Eventually they heard the door on the far end of the gym open and close and all was quiet again.

"Ok," said Dana. "I've been thinking about this thing we saw take over a whole room full of people. I call it the Scarecrow. It has some sort of mind control but it looked like it needed direct physical contact. Or at least it did until it took control of Professor Moreau and used his link with the lizards. I think the people here at this school already knew about this thing and are jamming the frequencies it uses.

"Prima and Lemma, you built this Faraday cage because you were experimenting with something similar and noticed the jamming whenever you started using those frequencies, correct?"

"That's right," said Lemma. "As soon as we started using a signal, something would flood it with a high-power random-noise broadcast. But look, that thing has my mother and sister. Do you think — "

"I think they will be ok if we can stop this thing," said Dana. "Bear with me for a minute. So Professor Moreau has these lizards that he uses to spy on people. I think he's using a similar mechanism to see through their eyes but he figured out a way around the signal jamming. And now that the Scarecrow has him..."

"It can control anyone if a lizard is riding on them," said Jayla.

"Exactly," said Dana. "So we need to figure out how the lizards are avoiding the jamming and see if we can change that. Any ideas?"

Everyone thought for a moment, and then Jayla said, "The usual way around jamming is frequency hopping. Which was first conceived of during World War II by Hedy Lamarr, who was also a very famous movie actress at the time."[100]

Tevi said, "That's right! It is essentially a cryptographic system..." Toma said, "...because both parties need secret knowledge about which frequency they will hop to next."

"Ok," said Dana, "So if we know the secret code of the frequency hops we can jam them. How do we figure that out?"

"Maybe we can ask a lizard?" said Noah. He pulled out the specimen bag he had stashed the captured lizard in earlier.

[100] U.S. Patent Number 2,292,387.

"Prima, go get the amplifier we use for the rat, and one of the receivers," Lemma said.

Prima groaned and said, "My leg hurts a lot still. Maybe you could do your own fetching and carrying for once?"

Lemma bristled. "Well, you don't have to be so..."

"I'll get them," said Jayla. She pulled herself out from under the table and lifted herself up to the edge with one arm while retrieving the helmet and a piece of radio equipment with the other.

What followed was a multi-hour radio hardware hacking session. There were a few granola bars in Lemma's backpack, and they had all recognized the fact that if they needed to use the bathroom they were just going to have to hold it.

Jayla knew the most about radio systems, and eventually even Lemma and Dana, who both were naturally inclined to try to take charge, deferred to her. Prima managed to supply information on how the rat helmet had been designed, and it became clear she had been the primary builder of the system, not Lemma. Jayla built something that worked like an active listening unit for a jammer, quickly moving through frequencies until it found a new signal. Noah figured out how to attach the rat helmet to the specimen case so it could read the lizard's signals without any chance that it would get away. The math kids worked on figuring out the pattern of the signal hopping. Dana and Lemma were the least useful to the process, and were resting at opposite ends of the table. It had gotten dark. Dana had found an extra bundle of the conductive fabric and was using it as a pillow, thinking they might have to spend the night there, when the problem was finally solved.

"I think we've got it. It's actually simple but amusing because..." said Toma, "...the hopping is based on the Fibonacci Sequence,[101] which..."

[101] The *Fibonacci sequence* is constructed by a simple rule: Add the two previous numbers to get the next number. Like this: 1, 1, 2, 3, 5, 8, 13, 21, 34, 55, 89... and so on.

said Tevi, "...is closely related to the golden ratio, Phi, which..." said Toma, "...is, of course, the symbol of the Science Discipline, of which..." said Tevi, "...Professor Moreau is Master," said Toma.

"Ok," said Dana. "So can we jam the signal?"

Jayla replied, "I think so. I can make this listening device scan for the signal and compute a match for the hopping. But I don't have enough broadcast power to cover the campus. Really, I think maybe a few people close together would be protected. It depends on how well the lizards can pick their signal out of extraneous noise."

"This lizard broadcasts a lot more strongly than any of the rats did," said Prima.

"Right," said Dana. "They're sending signals back to Professor Moreau. I think we've established that he built these lizards to be better at it than anything in nature." She looked at Prima. "Um, I have to ask. It's pretty clear that you did most of the work on this project. Why did you team up with Lemma?"

Lemma heard this and said, "It was my idea and I was involved all the way along. Prima is better with the electronics than I am. I figured out the biology."

"Is that true?" Dana asked Prima.

"Well, she did seem to know exactly where to look for the signals. I thought that maybe she was getting information from her mother and sister. But this is groundbreaking stuff, so I had no idea where they could be getting it from. It's definitely never been published."

"They didn't tell me what to do," said Lemma. "They told me when they thought I was on the right or wrong track. But I worked it all out on my own."

"Hmmm..." said Dana. "Anyway, none of that is important right now. Jayla, are we ready to test this lizard jammer?"

"I guess so. It's not going to be super portable. I only had this large laptop, that the science kids were using to display their results, to run the frequency-hopping algorithm. Its battery can also power the jammer. But you'll have to carry the whole thing and I don't know how long the battery will last."

"One thing at a time," said Dana. "We can worry about that after we see if it even works. Hmmm... Ok, here is the plan to test it. Grab the cage with the rats in it and move it out of the Faraday cage. Then we will put the lizard in and see if it makes them go zombie."

"Ok," said Noah. "But try not to let my specimen escape."

"How will we know if a rat turns zombie?" asked Lemma skeptically.

"Good question. Do you have any treats the rats really like?" She took the box of treats that Lemma offered her. "I'll give them a bunch of treats. If they stop eating them when the lizard touches them, I will assume zombie rats."

"Sounds good," said Noah.

"Ok, here I go!"

Dana poured some treats into the rat cage, then opened the specimen case and dropped ML5 in. She opened the Faraday cage door and stepped out, holding both the laptop, rat cage, and radio equipment.

"How does it look?" asked Noah.

"Yeah. I think it's good." The lizard was sitting on the back of one of the rats, but both rats were ignoring it and happily munching treats. Then the lizard scuttled away into the corner and attempted to hide under the wood shavings at the bottom of the cage. "Rats looked fine when touched, and the lizard isn't behaving like it's controlled either. It's hiding, not spying."

Dana stepped back inside the Faraday cage, closed the door, and joined the others back under the table.

"All right, we have a potentially useful tool. Now we need to figure out how we actually use it. Any ideas?"

They all sat thinking. No one had any immediate ideas.

Chip said, ***"Well, if no one else is going to state the obvious: The working jammer, that has very limited range, needs to be patched into the one that doesn't work but has enough power to cover the whole campus."***

Dana looked at Chip and said, “That does seem obvious now that you say it.”

“You do know that no one else can hear the other end of your delusional conversations with the magic fox, right?” said Lemma.

“Oops. Sorry. He says that we need to find the main campus signal jammer and attach this one that works, so we can protect the whole school.”

“Admittedly a good idea,” said Lemma. “The voices in your head are smarter than you are, apparently.”

“Maybe I am a little smarter than you but I am a whole lot smarter than her. You can tell her I said that!” said Chip.

Dana ignored both of them. “Ok, who wants to go with me?”

Again silence. Finally, Noah said, “We’ve been in it together so far. I’m with you.”

“I would go if I could,” said Prima, “but with my leg, I would probably slow you down.”

“Me too,” said Jayla, “but I do think there’s a way we can help from here. I’m going to set up a two-way radio and run the antenna outside the Faraday cage. If you need us for help connecting the equipment, we will be here. I’ll listen on our usual frequency.”

“And if you have any math problems...” said Toma, “...we’ll be here too,” said Tevi.

“I’ll stay here and watch over my lab partner,” said Lemma.

“That’s so good of you,” said Dana sarcastically. ”Come on, Noah, let’s go.” She started to slide out from under the table but then had another thought. She turned back and grabbed the bundle of the extra conductive cloth she had used briefly as a pillow.

“Never know when you may need a quick Faraday cage,” she said. “But this is now a lot of stuff to carry. Does anyone have a bag?”

“There’s a backpack over here...” said Toma, “...that looks perfect,” said Tevi.

“Hey! That’s my backpack,” said Lemma. Everyone — even Chip — turned to look at her with disbelief. “Which you can, of course, use...

since it's an emergency..." Dana loaded up the backpack and put it on. Then, followed by Noah and Chip, she got out from under the table and left the cage, carrying the bundle of cloth and the laptop lizard jammer.

"Ok, Chip," she said, "If you're my subconscious, where exactly am I thinking the jamming station is broadcasting from?"

"Weird green dome cupola, probably made of copper, on top of Myrtlegrove Manor," said Chip.

"Where are we going?" asked Noah, as he followed Dana towards the exit at the other end of the gym.

"The roof of the manor house," said Dana. "And we're going in style! Let's go get the Beetle."

When they reached the Beetle, they found that Bosco was no longer tied to the landing gear. His leash was still there, but he was gone.

Dana worried briefly that he had been taken by one of the zombies, but there was no time to look for him. They had already spent far too long hiding out and creating the jammer. Who knew what had gone on during that time with all the people the Scarecrow was controlling. Hopefully no one had been badly hurt, but that was probably going to happen eventually. Even saving Bosco probably depended on getting this new jammer installed as fast as possible.

The two children and the fox boarded the Beetle and launched into the dusk. As they flew towards the manor house, lizards started to land on the outside of their vehicle. They had been spotted. By the time they reached the manor house, there were a half dozen of the things on the windows.

They hovered near the strange green cupola and Noah took the controls while Dana donned the backpack. "Stay near, in case I need to get out in a hurry. I'm going in," she said to him.

She checked to make sure the hatch was clear of lizards. She was wearing the lizard jammer, so she needed to make sure none of them got inside the Beetle, or Noah and Chip would be unprotected. The coast was clear, so she exited the craft onto the landing plate and closed the outer airlock door behind her. A lizard landed on her and she felt a slight fuzziness in her head for a moment, but then the jammer kicked in. *Excellent! It's working.* Then another lizard was on her, then another. *Uh oh. What if multiple lizards can produce a combined control signal strong enough to overwhelm the jammer? I need to finish this, fast.*

Noah was holding the Beetle in position. She needed to slide over the green railing. But as she was about to execute that maneuver, the whole craft lurched violently sideways. Something a lot larger than a lizard had hit it.

She looked up and saw the albino chemistry teacher, Dr. Gretta Griffin, clinging to the side. In the low light, her eyes glowed red, and her teeth flashed white and... pointy. *Pointy*? Long canines protruded from her mouth.

"OH FOR THE LOVE OF HALLEY'S COMET, *VAMPIRES?!*" Dana shouted. Dana couldn't see the lizard on Zombie Griffin, but had no doubt it was there and that the Scarecrow was in complete control of her actions.

Even as Noah compensated for the extra weight, the Beetle was hit by a second large body. It was Dr. Griffin's husband Hanz, eyes glowing, white teeth flashing. "Of course! What could be better than one vampire? *Two* vampires!"

The zombie-vampires began to pound the craft with incredibly strong fists. Without being pressurized, the windows could not take the punishment. They began to cave in. It looked like Noah was panicking now, and the Beetle pitched and spun. For an instant, the green railing was lined up in front of her and Dana went for it.

She dove into the cupola as the Beetle went spinning off into the night, tipping to one side and accelerating diagonally downward. ***"I think this is probably gonna hurt,"*** said Chip, who was on the receding Beetle, but apparently still within whatever their range of communication might be.

Dana felt sick for fear that Noah and Chip would be hurt by the crash or *zombie vampire* attack, but she knew the best way to help him was to keep going and get the lizard jammer installed fast. She looked up at the domed roof above her, and spotted a wire attached to it. *Yes! This has to be it.* She followed the wire with her eyes. It went down the outside of cupola and ran along a short railed walkway towards a small door into the attic of the manor house. As Dana started towards the small door, two more flying lizards landed on her. She made an effort to brush them away, but she felt like the fuzziness in her head was definitely increasing with each new lizard that landed on her.

She opened the door and got inside, closing it behind her, finding the latch, and locking it. Once inside, she searched her body, removing lizards. She was careful, as she didn't like to hurt any living thing, even if this might be a clear case of self-defense. But it wasn't the lizards' fault they were being controlled by some sort of monster.

There was a stairway, and she followed the wire, tacked along the top of the crown molding, down the stairs and to the end of the hall where it disappeared through the wall above a door. She opened the door and was relieved to discover exactly what she was looking for: a small room containing some sort of radio equipment setup.

Dana pulled the ham radio handset out of the backpack to contact Jayla and describe the setup. But when she turned it on, before she could even query for her, Jayla's voice was on the line. "I don't know if you can hear me, Dana, but they found us!"

"Jayla! I'm here."

"Oh no, it's Master Skadi!" There was a crashing sound over the radio, then it went dead.

"Jayla! Jayla!" but there was no answer. "Jayla, come in!" she yelled.

Only belatedly did she realize that she was making far too much noise inside what was now enemy territory. When she heard something at the door behind her, she reflexively jumped to one side. The Master of Engineering was in the process of leveling the glowing slide rule at her. The green beam from the slide rule flashed out, missing Dana but hitting the radio equipment. The radio exploded in sparks and flames, a hole punched through it and also through the wall behind it.

Dana acted without thought, pushing off the wall in a fast low dive towards the tiny woman before she could aim and fire again. Dana swung the backpack and knocked the slide rule up and aside. Another deadly beam fired into the ceiling and, before she could re-aim, Dana punched her hard, squarely in the face. It was either Dana's punch, or the lizard jammer's proximity cutting the Scarecrow's link, but the professor collapsed, unconscious.

Dana breathed a sigh of relief. She found the lizard in the engineer's hair and removed it. Then she looked back at the destroyed radio equipment in horror, as the implications of what had happened dawned on her. Not only would she be unable to amplify the frequency-hopping jammer that Jayla and Prima had cobbled together, but the more powerful reactive jammer was now a smoldering wreck. It seemed likely that the Scarecrow would no longer need Professor Moreau and the lizards to control people from a distance.

Dana was the only one with a personal portable jammer. Everyone else within the Scarecrow's range, whatever that range might be, was quite likely now under its direct control.

CHAPTER 35 — WRAP UP

I'm all alone. It's all up to me.

There was a crash in the attic above her. The little door that led out to the green cupola on the roof had been forced open. *I'm not alone any more. Albino-Zombie-Vampires!*

Dana hurried down the hall to where the stairs continued. She descended as quickly as she could, leaping several stairs at a time, but was only a few flights ahead when she heard her pursuit on the stairs above her. By the time she reached the familiar second floor, they were even closer.

She recognized the corridor that the bottom of the stairway opened into, and bounce-ran down it towards the main sweeping staircase to the entry hallway. She had no plan of action if the main zombie mob was still there. She hoped they had moved on to some other location.

They had not.

As she came around the sweeping staircase, she saw that everything was much as it had been hours before. The zombies were huddled together in the remains of the original arrangement of chairs that had been set up for the contest winners announcement. Zombie Moreau was in the middle with the ball of rubbish that was the Scarecrow still riding his back, a single glowing green eye peeking over his shoulder. She could see familiar faces in the mob, staring blankly. Robot Cavor's bisected broken body still lay by the open entryway doors. Dana wanted to turn back at the sight, but she heard the pursuers behind her reach the hallway, and knew they would be on her in a moment.

Seeing her, the zombie throng, all still maintaining contact, swept up the stairs towards her. Scarecrow-Moreau stayed in the back of the pack, while Zombie Uachen was in the lead. His hands were no longer glowing and he didn't move with the same grace and precision that Dana had witnessed earlier, but seeing him coming, Dana still knew that she had made a mistake. As scary as the vampires were, they were not directly linked to the Scarecrow. Her jammer was capable of cutting the signal to them, if she got close enough. But the control exerted through these zombies, with a flesh-to-flesh chain connecting them to the Scarecrow, passed through internal bodily electric fields, and could likely not be so easily jammed. She momentarily considered turning back, but an old adage flashed through her mind: *She who hesitates is lost.*

Dana kicked off her shoes and hopped up onto the banister of the long curving staircase in her socks. She started out sliding slowly, due to her low gravity, but she picked up speed as she moved.

Zombie Uachen, a kite tail of other bodies in tow, moved to the side of the stairs to reach out for her, but she hopped over his grab. She continued to accelerate downward. At the last moment, before reaching the end of the railing, she crouched and jumped hard, launching herself into the air and across the room, straight for Zombie Moreau.

"Kiai-yah!" she whooped. In the air, her right leg flashed out in a flying *mae geri* kick to the head of her real opponent: the facelike ball of rags and trash riding Zombie Moreau over his left shoulder. She felt a flash of pressure of its attempt to control her mind. A sense of incompleteness rushed through her, a shattered tiny piece of something that longed to be greater — wanted to be made whole. Fortunately, the contact was too brief for her to be overwhelmed, and the sensation vanished instantly as her kick sent the thing flying backwards across the room. It ended up in the corner, striking the wall hard.

Dana passed over Zombie Moreau, landing on her feet, and continuing to slide in the direction of the thing scrabbling in the corner. Her socks were nearly as frictionless on the polished floor as they had been on the bannister. She was torn: she had to get to the Scarecrow for her jammer to cut contact to all its victims, but, if she actually touched it for more than an instant, it would surely take control of her. There was no time to think;

it was already moving, on tiny clawed paws, back across the floor towards her.

"Use the tablecloth!" Chip was there, standing in the open doorway.

"Got it!" Dana reached over her shoulder, into her partially unzipped backpack, and pulled out the folded bundle of electrically conductive cloth. She gripped one edge and snapped it outward so it unfurled over the Scarecrow, settling down before the strange creature could get out from under the edge. Then she dropped to her knees and scooped the horrible thing up and rolled it in layer after layer of a makeshift combination of Faraday cage and monster-binding cocoon.

All of the zombies behind her fell to the floor, unconscious but free. The two zombie vampires, freed from control in mid-leap, crashed, fully asleep, onto the floor on either side of her, sliding to a stop against the wall.

"Good job, kid," said Chip.

"You too, kit,[102]" said Dana.

Dana sat still for a moment, panting, letting the tension drain out of her. The bundle in her arms stopped struggling. She didn't know if she had smothered it or if it was just pausing to regroup. Part of her hoped she hadn't killed it, whatever it was. Another part of her didn't care — she wouldn't mind if it was gone forever.

She looked around the room, trying to decide what she needed to do. This was a real mess. She was party to a big secret: the Lunar Society and its advanced technology. Not all the people who had been here were part of that secret. How could this be kept covered up? How much would those who had been zombies remember?

Her eyes fell on the destroyed machine that had recently been Dr. Cavor. Then she looked at the collapsed form of Professor Moreau. He was on his back, snoring loudly now. And as she watched, a lizard crawled out of his mouth and gave a little yawn. She shuddered, but realized that

---

[102] A *kit* is a young fox.

if anyone could fix this mess, it was him. Noah was right; he was second in command here. So she overcame her revulsion and walked over to him.

She bent down, shook his shoulder, and said, "Professor Moreau, please wake up," but he continued to snore. She slapped his face lightly. Then harder. "Wake up!" she yelled into his ear.

Finally he sputtered and blinked his eyes, "Hrmm urmm bleh... What is going on?" he mumbled groggily.

"We were attacked," said Dana. "The whole school was attacked. Something that takes over people's minds. I've been calling it the Scarecrow."

He was coming fully awake now, sitting up and taking in the scene around him. "I know of it," he said. Then he saw what was left of Dr. Cavor. "Catherine! How? It never had that kind of power. Not in hundreds of years..."

"Once it took over your mind, it used your lizards to take others. Dr. Taxidiótis did that to Dr. Cavor with some sort of death ray. Did you know she was a robot?"

"A telepresence[103] unit, yes," said Professor Moreau. "Or at least it used to be. I used to see her every year or so, in the flesh. But it has been a long time now..."

"I know you went through some stuff, but I need you on your feet and giving orders. She is down, so you are in charge. We need to clean up this mess. I assume that not all of those people in that heap over there have any idea what goes on at this school. I'm not sure I know either, but I know a little, and I know it's supposed to stay secret."

"Right," he said. "Who else is still functional?" He rose to his feet.

"It's just you and me. I was the last one standing."

"It took over everyone else and then you beat them all by yourself?"

---

[103] *Telepresence* refers to technology which allows a person to feel as if they were present, to give the appearance of being present, and to have an effect at a place other than their true location.

he asked, incredulous. Then looked at her and said. "Oh, of course you did. You are — "

"Noah helped... Oh! No! Noah!" She ran out the front door without another word, still wearing the backpack jamming rig and holding the wrapped up Scarecrow-head-creature in her arms.

Dana circled the mansion until she saw the Beetle. It was almost fully dark outside now, but the grounds lights had come on, and she spotted the crashed craft. She gasped at the state it was in and bounce-ran faster, searching the wreckage for Noah and fearing the worst. The little spacecraft had smashed hard into the trunk of the big oak tree in front of the manor house. Between the vampires and the collision, the vehicle had been torn apart. The outer shell of the Beetle was cracked like an egg, with the Beetle-heart protruding from the twisted remains of the outer shell.

The gravity shutters were in tatters, some of them detached completely, scattered and spinning and chaotically flopping about like injured metal moths. Dana remembered the first trick she had done with the glued-together drink coaster. As she ran up under the wreckage, one of the detached gravity-shielding slats came spinning out of the darkness like a thrown axe. Dana barely dodged out of its way. It hit the trunk of the oak tree and ricocheted off at a different angle, now fluttering like a large piece of confetti in ticker-tape parade. *Of course, the coasters never tried to kill me.*

Without effective gravity shielding, the tanks of neg-grav water in the core were being rejected by the Earth. The whole structure would have been pulled skyward, except it had gotten caught in the branches of the oak tree. Dana could hear branches creaking and cracking under the tremendous neg-grav force of the water in the tanks. As she neared the tree, she could see the whole broken mess, illuminated by one of the landscape spotlights, shifting and moving upward in jerks and fits.

Some of the water tanks were cracked and leaking. Neg-grav water was trickling up through the branches before dripping upward into the night. Any second, the whole craft would heave its way through those branches and be lost to the endless sky.

"Noah!" Dana yelled for her friend. "Noah, where are you?"

No response. She scanned the scene of the crash. Then she spotted a shoe and part of a leg, hanging out of a partial section of the Beetle, attached to the whole wrecked mess. He was still inside the demolished craft, maybe ten feet above her head.

She started to think about the best way to get to him, first looking for a place she could put down the bundle wrapped in her arms, something she could pin it under so if the thing inside was still alive, it couldn't escape the cocoon. *Should have left it with Professor Moreau*, she thought.

CRACKKKK!

A thick branch had given up, and the wrecked Beetle slid maybe another couple of meters towards the sky, taking Noah with it. Dana stopped thinking and let her instincts take over. She ran at the trunk of the tree and up it, pushing off in an upward leap as she rebounded, trying to gain the necessary altitude to reach the lower edge of the wreckage of the poor broken little spacecraft.

Her acrobatics and currently lowered gravity profile were enough to do the trick. She caught hold of the lowest hanging piece of aluminum tubing with one hand, still holding the bundled Scarecrow under her other arm. She pulled herself up into the bowl, formed by three pentagonal polycarbonate faces, where Noah lay unconscious. Noah's shirt tail was caught on part of the Beetle's broken frame, and she began to tug on it, trying to free him. Unfortunately, her weight and motion rocked the wreckage and gave it the motion it needed to slide around one of the tree's major branches.

With more snapping and crashing, the Beetle started moving with increasing speed upward through the tree, into even thinner branches. The crowbar was there, next to Noah, so she set the Scarecrow down, picked the bar up, and pried at the aluminum frame, trying desperately to free her unconscious friend.

Another spinning, fluttering gravity shutter came her way and she knocked it aside with the crowbar without thinking. Then she noticed another movement. The bunched-up tangle of conductive fabric was climbing up the framework towards the tanks. The creature had chewed and clawed its way through the fabric, and enough of its head was visible for Dana to finally see its face. She was surprised to see that it was an opossum. It hissed at her, and she saw it only had one bright green eye. The place where the other eye had been was perfectly smooth and fur-covered — no eye or eye socket or trace of a scar. Just like Chip!

There was no more time. The remains of the Beetle were free of the tree and rising, carrying them all upwards. She could either grab and rewrap the Scarecrow-Possum, or save Noah, not both. She barely hesitated.

She dropped the crowbar, grabbed Noah's limp body, and hauled on him as hard as she possibly could. Clothing ripped, aluminum bent, and Noah came suddenly free. Dana tumbled backwards out of what was left of the Beetle, holding Noah tightly in her arms. They fell backward, fairly slowly, down to the ground below. Dana crashed down onto her back, holding Noah protectively. She heard the electronics in her backpack go *crunch.*

Then she felt her head go all fuzzy again, as it had earlier during momentary direct contact with the Scarecrow, and she expected to lose herself to its control. But the feeling faded as the broken Beetle rose higher and higher, beyond the effective range of its mind control trick.

She lay there physically and mentally drained, as her very first spaceship headed back into outer space without her. "Goodbye," she whispered.

"Mrrm hururm..." Noah moved in her arms. "What happened?" he asked.

Relief for her friend flooded into Dana, displacing any regret at the loss of the Beetle or the escape of the Scarecrow. It would probably die in space anyway. How could it possibly survive?

"Noah! Are you ok? What is the last thing you remember?" she asked her newly-awakened friend.

"We were on our way out of the gym. We were going to try to patch the new jammer into the old jamming broadcast. Did we win?"

"Did we win? Did we win the *contest?*" She realized she didn't know. Dr. Cavor hadn't yet declared the winner when the Scarecrow came in. It hadn't even occurred to her since, but it was still an open question. Then she realized he was talking about beating the horrible Scarecrow, not the somewhat less horrible Lemma.

"Yes. We beat the monster. But Dr. Cavor is still dead. I mean, still dead again. Well, maybe. I think we will probably be seeing more of her. Unless she was actually already dead before now..."

"What? I just woke up, but I'm pretty sure its you who isn't making sense right now."

"It's complicated. She was a robot. But maybe piloted. But Professor Moreau says he hasn't seen the real flesh-and-blood Dr. Cavor in a long time. So maybe a robot?"

"I would ask for more clarification, but I don't think I'm up to it yet." He groaned and stretched. "I ache all over. What happened?"

"Most recently, a couple of vampires ripped apart the Beetle and you went out of control and crashed."

"Zombies AND vampires?! What, were there werewolves too?"

"Right, you know those albino teachers..." Just then Dana's phone started to ring. "Look, we'll go over everything later, ok?" She didn't wait for him to answer and pulled her phone out of her pocket.

Her mother was calling.

CHAPTER 36 — DANA MA[104]

Dana finished up with "Ok Ma, I love you too!" She hung up the phone and returned her attention to Noah. "She's at the airport. She just got her car and will be here to pick us up shortly."

"You didn't tell her anything!" said Noah.

"What was I going to say? What would she believe?"

Noah stood up, still a bit dazed. He looked down at the Beetle's remnants scattered at the base of the tree, and then looked up. "The Beetle's gone?" he asked, gazing at the early evening sky.

"Yeah," said Dana standing up. "It's gone. And the Scarecrow."

"Whoa. I guess you're gonna have to give me the whole story later. Where is everyone else? Are they ok?" He looked down and nudged at a twisted bit of aluminum frame with his foot. "Hey," he said. "Where are your shoes?"

Dana looked at her feet. She had a hole in one sock and her little toe was poking through. "Oh right. You know how you're always saying I'm gonna get myself killed doing crazy acrobatic stunts? Well, you should have seen the way I took out the Scarecrow! But come on — right now we need to find the others and see if they're ok. The last thing I heard over the radio was Zombie Skadi."

Noah started to tug at his ruined shirt hem, obviously concerned at this news. He looked towards the gym, peering across the dark campus. "Look! — there they are," he exclaimed.

---

[104] *Denouement* /dā-nü-mä^{n}/ means "the final part of a play, movie, or narrative in which the strands of the plot are drawn together and all the remaining issues are explained or resolved." (Unless you are planning on writing one or more sequels!)

Dana saw Master Skadi and the others emerging into the lamplight. With a sigh of relief, she could see that Skadi was regular Master Skadi, not Zombie Skadi. She was carrying Prima, cradled in her arms like a baby. Jayla, Lemma, and the twins all appeared unharmed.

Dana and Noah rushed over to meet them. "You're all ok!" Noah said.

"I heard you on the radio. I thought Zombie Skadi got you!" said Dana.

Jayla said, "No! We got her. Or Lemma did. She was brilliant. But the radio got smashed."

"How did you get her?" asked Dana.

"When she came in for us, Lemma snuck around her and closed the door to the Faraday cage," said Prima. "Master Skadi passed out immediately but then she fell on the radio."

"It was nothing," said Lemma. "I mean, of course I was amazing, but I'm sure you would have done the same."

"How did you know it was safe to come back?" said Dana.

"Master Skadi woke up after she fell into the Faraday cage," said Tevi. "Then we tested the lizard again, and," said Toma. "and the rats weren't zombies," said Tevi. "So we came back for you," said Toma. "We were worried," said Tevi. Toma nodded.

"We're ok," said Dana.

"Come, young ones. We must see to those inside," Master Skadi said. Dana noticed that she had streams of dried blood running from beneath her eyepatch.

"Master Skadi, are *you* ok? What happened to your face?"

"Yes, brave Dana. I am... fine."

Tevi chimed in. "The Scarecrow took her eye. "No," contradicted Toma. "When she woke up she said it was her *purpose* that was gone, not her eye," said Toma.

"What did you have in your eye?" asked Dana, somewhat alarmed. *Chip had a stone in place of his eye, and the possum certainly did too.* Dana thought back to her first glimpse of the deer outside her house. It had had the possum riding it then, she now realized. *Did it have a stone the whole time?*

Master Skadi shook her head and began walking to the manor steps. "Come, children. We must provide aid to the others."

Lemma piped up. "Yeah — we need to get Prima some help."

"I'm ok," Prima said from Master Skadi's arms. "I can walk a little. Master Skadi only carried me the whole way because I didn't want to slow everybody down. Master Skadi, I can probably get down now. Thank you for the ride."

Master Skadi smiled, and the dried blood on her cheek cracked a little. "You are most welcome, child." She gently set Prima back on her feet.

They walked up the ramp to the manor entrance. Prima hobbled along, grimacing a little but able to keep stride. Suddenly, there was a commotion inside the entryway, and the group froze. A small herd of deer — four, to be exact: a buck, two does, and a small spotted fawn — burst forth from the manor house and sprinted across the campus lawn, bounding so quickly it almost seemed like they were in partial gravity.

No more deeripede, she thought, and was glad no animals were harmed in the making and unmaking of the Scarecrow. The group entered the manor house. All the teachers were awake and whispering in a circle, including Dr. Taxidiótis, who had came down from the upstairs hallway where Dana had last seen her. She was the first to look up and see the students and Master Skadi enter, and Dana could see that her sucker-punched eye was red and beginning to swell shut. She nudged Professor Moreau and pointed at the doorway. Professor Moreau put a finger to his lips to indicate they should keep quiet, and he motioned them over. All the parents and other people who had come to see the announcement were still out cold, except for Lambda and her mother. Lemma ran to them and threw her arms around her sister. Jayla, Prima, and Toma and Tevi moved to find their own unconscious families, but Professor Moreau stopped them.

"Wait!" he said in a hushed voice. "We have things of importance to discuss." The teachers' circle parted for Master Skadi and the STEM contest finalists to step inside. Dana had noticed that both halves of Dr. Cavor's broken robotic body had disappeared. It didn't take a genius to figure out that a coverup was brewing.

"Before anybody is awakened, we must agree on a narrative," said Professor Moreau. "We must get our stories straight so they do not contradict one another. But first, are any of you hurt?"

Professor Moreau looked at each of children appraising them for signs of injury. They all shook their heads, even Prima.

"Prima hurt her leg," said Toma. "In the fight with Master Skadi," said Tevi. "While she was still a zombie," said Toma. "While Master Skadi was still a zombie, not Prima," Tevi said. "Prima was never a zombie," agreed Toma.

Professor Moreau looked at Master Skadi, who nodded. "She fought well," she said. "But she was injured in the battle."

"I'm fine, everyone. Really. See?" Prima stood upright from where she had been leaning on Toma for support, and winced. "Mostly fine."

"We're all mostly fine," said Dana. "But if you want our cooperation, with the story that gets told about what happened here, there are some things we need to know. Right?" Dana looked at the other children. They all nodded, including Lemma and Jayla.

"No one would believe the truth anyway," said Jayla. "It's just too... too..." She shook her head and didn't finish the sentence. Dana was relieved to see that Jayla seemed to understand the importance of keeping the school's secrets, if only because some things are too impossible to explain.

Professor Moreau nodded. "You all acted courageously, and after what you have seen, that is only fair. We do not have much time, but I will do my best to help you understand."

Dana looked at the Lei family. Lemma still had her arms wrapped tight around her sister. "Why are Lemma's mom and sister awake and no one else's families are? Mine and Noah's parents aren't here, but look how worried everyone is!"

Jayla was tugging at her braids with both hands and scanning the room, clearly anxious about her grandmother. Prima was holding onto Toma's shoulder for support again, and looking very much like she wanted to run into the collapsed crowd to find her parents, knee injury notwithstanding. The twins were grasping each other's hands so tightly their knuckles had turned white.

"Now listen here! I'll have you know — " Monera Lei started, but Professor Moreau interrupted.

"Monera, please. They are concerned for their loved ones, as you were," Professor Moreau said. He turned back to the children. "Let me first assure you all that your families will be fine. They will wake up a little sore, without memory of the events, but there does not seem to be any lingering effects of herm done. As for the Leis, they are a very old and respected Lunar family, and trusted allies. Monera will provide valuable assistance in managing this crisis."

"Ugh, whatever," said Dana. "Next question. I found a class list for the daycare here from a half dozen years ago. Every one of us in the STEM contest was on that list, except for Noah. How is that possible? How are we all genius kids? Did you do something to us?"

"That was Dr. Cavor's project, so that would be a question for her. However, while I could be wrong, I do not think anything was done to you, other than teach you. But you may have been taught some unusual things. You are all surprising in more ways than you probably know. For example, having experienced what you have, most people would be in complete shock. You children acted courageously and with a great deal of calm under extreme pressure. I believe that Catherine taught you all some very advanced coping mechanisms at a very early age, so much so that you may not even realize what gifts you have. That is, all of you except for Noah."

Professor Moreau's voice took on a kind tone, and the faintest hint of a smile tugged at the corner of his lips. "Noah, my boy, you have some natural gifts of your own."

Noah opened his mouth as if to ask a question, but then stopped. He let go of his shirt hem and looked at Professor Moreau, thoughtfully.

Professor Moreau continued. "As to how she was able to choose so many young geniuses... well, only she knows the answer. But there is a prophecy... I hate to use the word 'prophecy' because it sounds painfully unscientific. A lunar scientist built a device once, long ago. Something that bent light far enough to make it go backwards. She could use it to see into the past and future. Some doubt the experiment because it has not been replicated. But she recorded seeing humanity protected from the Xenos in the past by a warrior who had the title 'Girl on the Moon' and seeing the reincarnation of that leader as a child in the early twenty-first century. That the girl would defend us again after making a trip to the Moon.

"Catherine believed that she knew who that child was, or at least, she had a set of candidates. Not everyone in the Lunar Society believes in this prophecy, but Catherine did."

"And you were spying on us with your lizards because she picked us for the Moonbeams Daycare. Because that made us the likely candidates?" Dana asked.

"Yes, in essence," he said.

Noah tugged at Dana's sleeve and pointed to the jumble of the recently zombified audience members. Some were beginning to stir and groan.

Professor Moreau saw this too. "If you have other questions, ask quickly. People will be waking up on their own soon."

"What about the stones?" asked Dana.

"What stones?"

"The stones from the moon. The one in Chip. And Dr. Cavor's umbrella."

"Ah. Not stones. Something else. But I do not know much more than you do. Catherine did not always share her knowledge with me, but I believe she had some theories and some evidence. She has been studying these for quite some time. She had discovered a way to utilize the one in her umbrella somehow. The Scarecrow, as you call it, it is indeed also connected. It has always been drawn to her umbrella when she experimented with it, so she has tried to keep it closely guarded. After this situation has been handled..." Professor Moreau paused to survey the scene, shaking his head. "After this has been dealt with, my priority will be to investigate what Catherine knew. What happened here is... unprecedented."

Professor Moreau does't know about the stone in the Scarecrow-possum's eye, Dana thought. *The Scarecrow was like Chip, somehow. Like Master Skadi? Maybe. She'd lost something from under her eye patch. Toma said she'd called it 'her purpose'. How much about the stones does Professor Moreau even know?* Dana considered that she herself might know more about this than anybody else except for maybe Dr. Cavor.

"And Dr. Cavor? What about her? Does she have another telepresence robot thing or whatever? Or is she really... this time is she really..." Dana trailed off.

Professor Moreau's face softened. "I do not have a good answer to that question. Catherine's activities are somewhat of a mystery to me. She knows things I do not. And she is not easy to keep watch over. She does not much care for my lizards, and..."

"Oh! I almost forgot. The moon bugs! Noah and I decided you were the one to tell — they killed Dr. Cav... or, well, she killed... anyway, I think the Moon Bugs may be a danger to us. To everyone."

"Yes, yes. I know. I saw. At least one of my lizards made it back from your trip, so in essence I was there with you on the Moon. The threat they pose will be addressed at a new full Moon Council."

"A New Full Moon?" asked Dana, confused.

"Ah, apologies. A new gathering of the entire Lunar Society Council. It has not happened in some time."

"*I* have a question," said Noah. "Your lizards. All the things they do by command. They even self-destruct, burning up. Did you do all that? Engineer the genes to make it happen?"

"Oh my, no," said Professor Moreau. "I was making use of existing genetic code, the way the thing that attacked us tonight did. That code is ancient, and quite unlikely to have evolved naturally. It is likely an ancient construct that was inserted into our ancestral genome. Much of what you might have been taught is 'junk' DNA... I am willing to discuss it at a later time, but our time here is running short." Dana could hear sirens now.

"Someone called the police?" Dana asked.

"I did," said Monera Lei. "These people who are not part of the Lunar Society," she pointed to those still unconscious, "will have lost a lot of time. But I decided we are still within the window of some plausible explanation if we act quickly. I am the Lunar Council Solar Relations Officer, I decide how we handle the Sunbeams."

"Sunbeams?" asked Dana.

"Solars. People not of the Lunar Society. Regular people." said Professor Moreau. "Monera is our public relations officer, more or less."

"Ok, one last question" said Dana. "Maybe the most important one: Who won the STEM contest?"

The other kids all laughed. But they also looked at Professor Moreau expectantly.

"Catherine was the one who had the results of the voting," he said. "I do not know what she was going to say. And since we cannot retake the vote — " Professor Moreau surveyed the chaos. "— I suppose the four of us will need to decide." He nodded his head at the Masters of the other Disciplines and Master Skadi. "My feeling is that the most surprising and revolutionary accomplishment was building a spacecraft and flying to the Moon."

Dana's heart beat fast. Did that mean they had won?

"But that wasn't the Delta Team's proposed project!" protested Monera Lei.

"Oh yeah?" Dana responded. "Well, *I* happen to know that you helped Lemma with her project, and now I know where you got the advanced knowledge to be able to feed her information. Also, you had Lambda sabotage our system."

"That is a serious charge," said Professor Moreau. "Do you have proof?"

Dana considered that, and a thought struck her. "Maybe I do," she said. "We had a camera in our lab. It was set up to watch a lizard cage, but maybe..." She pulled out her phone and pulled up the online video record from the camera, fast-scrolling through it. "Let's see, it would have had to have been at about the time..." Suddenly she saw something and stopped, shocked. "Dr. Cavor?" People peered over her shoulder as she watched. On the video, the Beetle could be seen in the background, through a transparent aquarium. Dr. Cavor slid under and up inside the Beetle and could be seen working on the wiring.

"Why would Dr. Cavor...?"

Professor Moreau coughed politely. "It looks like Catherine had her own agenda and her belief may have made this prophecy self-fulfilling. Anyway, with no way to get at her data or reconstruct it, the judges will have to make a decision. I processed the data on one of my recorders earlier..." Dana had a brief flashback to seeing Professor Moreau swallowing lizards to "process" them and felt just a little bit queasy. "... and saw what happened going to and from the Moon. It is my opinion that Dana and Noah would never have made it back without the help of every other team here. In fact, you all acted as a collective team to get Dana and Noah home safely. So, I suggest that all the contestants be awarded the win as a single team. What do you say?" He turned to the other three judges and they all nodded, although it seemed to Dana more a decision of expediency than judiciousness.

"Then that is what we will do. It is unprecedented, but I do not suppose any of you should have reason to complain that Dr. Cavor saved the day for most of you by losing the electronic vote count when her robot body was destroyed. That is, unless, you are not a fan of the *deus ex machina* ending.[105]" He laughed at his own joke.

[105] *Deus ex machina* is the Latin translation of an ancient Greek phrase meaning "god from the machine." It refers to an unexpected power or event saving a seemingly hopeless situation, especially as a contrived plot device in a play or novel. In some ancient Greek plays, "gods" were lowered on ropes and pulleys at the end, suddenly appearing to resolve all remaining issues. Handy, if you can use it.

A scream rang out through the great hall. It was Mayor McCaws. He had sat up and was trying to rouse his unconscious son.

"Our time is up," said Professor Moreau. "Our official story is a natural gas leak. Udo and Tempora, open all the windows and doors, then wake people up and take them outside to get away from the gas."

Master Skadi said, "There is the nest of a grand bird in a tree outside. I will use it to block the chimney. Then I will turn on one of the heating controls, so the heat is running without purpose."

"Yes!" said Professor Moreau. "Blocked chimney... thermostat... very good. Do that. Gretta and Hanz, take the children back to their dorms. The parents can come get them, but the paramedics need to look over our 'gas leak' victims first. They should not have any memory of events. Officially, the children were not here and were never in any danger. Prima, you will need to invent a very ordinary reason how you hurt your leg."

Prima looked at Lemma, who nodded. "I was so excited about winning the contest that I... tripped and fell. My parents will believe that. It's happened before."

Professor Moreau nodded, then made a beeline for the mayor. "Mayor, please, calm yourself — such a display is unnecessary. The situation is under control. The police are almost here. Everyone is safe."

Before Dr. Griffin and Professor Hetzler led the children away from the scene, Dana remembered something. She bounded up the grand staircase and rescued her shoes, putting them on as she hopped back down the stairs. Of all the things she would need to explain to her parents, what had happened to her shoes might have been the most difficult.

The two albino teachers looked more normal now, but as they ushered the children out the front door, Dana couldn't help picture them as they had been on top of the cupola, all vamped out. Not to mention the strength they had exhibited, leaping through the air onto the Beetle and breaking welds with their fists and bare hands. It had been ripped apart. She was the only one who remembered seeing them that way. It sent a chill down her spine. She wasn't sure she could go back to the dorm with those monsters, as non-threatening as they might look now.

Fortunately, she didn't have to.

"Dana! What happened?" Dana's mother called out to them, running up from the parking area. Sirens could already be heard in the distance, getting closer.

"Mom!" Dana yelled and ran to her, long overdue tears welling up in her eyes, as Mary Dash embraced her daughter. She wanted to tell her mother everything that had happened. Instead she mumbled, "Gas leak. Or something. People just started passing out. We're fine."

"My goodness!" said Mom. "Is everyone ok? Is Noah with you... Oh, yes of course, you're right here, you two are inseparable. Come with me. Let's get you both out of here."

"Sounds good to me," said Noah, "but we need to get our stuff from our rooms first."

Dana realized that if they were leaving the school tonight, she needed some things from the lab too.

"Mom, could you go with Noah and pack up my stuff please? I need to run down to our lab. We left it a mess, and we want the school to think we're responsible students who clean up after ourselves."

"Yes, of course you do," said Mary Dash, "but be quick about it."

Dana headed towards the lab, waiting until she was around the side of the manor house and out of sight to break into her moon glide sprint. When she entered their lab for the first time in a week, she stopped and considered things. She really wanted to take the flying basket with her. What would happen to it if she left it here? Someone was sure to take it. But it wasn't something that would fit in the car, and how would she explain it to her mom? Where could she put it?

"Put it in the secret passage," said Chip as he wandered in behind her.

Dana nodded and headed over to move the bookcase. That done, she maneuvered the flying basket over and hauled it down and through the door. Then she looked around the room. Anything neg-grav went through the secret door. Flying coffee cup, weights and scale, rolled-up grav-trampoline. Everything. She closed the door and pushed the bookcase back into position.

Coverup complete, she grabbed all of her and Noah's stuff and got ready to leave.

"Aren't you forgetting something?" asked Chip.

"Hmmm? What?"

"Unless you want to explain my miraculous healing to your mom, you should bandage me up and carry me back in a cage."

That took a few more minutes to accomplish. Dana found a spare animal cage and some gauze on the equipment shelves, gave Chip a bandage eyepatch covering the missing eye, and put him in the cage. Then, carrying the fox cage and a few miscellaneous personal items, she headed for the dorms to meet Noah and her mom.

At the dorm rooms, everyone was packing up to go home.

"Oh, I forgot you had Dad's fox with you," said Mrs. Dash.

"Yeah. He was no trouble. And I think he's getting a lot better. We've even had him out of his cage and were teaching him tricks."

"Where's Bosco? Did he give you any trouble?"

"Not at all," said Dana smiling. "In fact, we were really glad to have him around. I'm sure he'll turn up, you know how he disappears from home all the time."

"But he doesn't know this area — what if he left and gets lost?"

Dana shook her head. "I don't see him going anywhere without me," she said. Mrs. Dash looked skeptical, but said nothing.

On the way out, some of the other kids were in the common area, waiting to be collected. Dana hugged Toma and Tevi together and Noah hugged Prima. They all promised they would write and said they looked forward to seeing each other at school in the fall. Then Jayla wheeled out of her room, followed by her grandmother pulling a rolling case. Dana ran over to her.

She bent down and embraced her new friend and said, "It feels like we have to say goodbye so fast. I'll miss you most of all!"

"I'll miss you too!" said Jayla. "But I'll see you in a month! We're going to have such a great time. I've never been so excited for the first day of school!"

As they headed back down the hill to the car, Bosco rejoined them, trotting up alongside Dana as nonchalantly as a dog could. The circular drive of the manor house was full of emergency vehicles, lights flashing. Mrs. Dash was stopped by a paramedic, concerned that the kids had been involved in the incident.

"The children are fine, but they're very tired and we'd all desperately like to go home, if we could just please get by you..." Dana noticed Professor Moreau was on the entrance steps, having an animated conversation with Principal Peters and Mayor McCaws. Matt was there too, hiding behind his father.

"...COULD HAVE BEEN KILLED!" Principal Peters was bellowing.

"Now, now, Pete," said the mayor. "This is indeed a serious, very serious situation, but one mustn't yell, mustn't yell. Professor Moreau, I'm very much afraid I must insist, must *insist* on talking to Dr. Cavor. We're going to need the building inspector to review your certificate of occupancy."

"And I am very much afraid that Dr. Cavor is on a sabbatical," Professor Moreau responded. "But I assure you I can handle any business in her — "

"SABBATICAL! SHE WAS HERE EARLIER!" Principal Peters apparently was still unable to control his volume. The mayor winced.

Before the mayor could speak again, Professor Moreau interjected, "I do not believe we need to involve the building inspector. We have already identified the issue. A bird built a nest in the chimney. Of course we always check for such things every autumn. But it seems someone was playing with the thermostat, which was turned on and all the way up."

"Still, still," the mayor said, "such an incident will require — "

"HER!" Principal Peters had spotted Dana watching them and was pointing right at her. "THAT DASH GIRL! SHE HAS A RECORD!" The volume of his voice finally dropped but Dana could still hear him clearly. "She has caused disaster scenes like this before. Always doing something dangerous. She probably turned on the thermostat. I would..."

"As a matter of fact," Professor Moreau interrupted him, "Ms. Dash was nowhere near the vicinity. But I believe that I saw this young man — Matthew, is it? — next to the thermostat in question."

"What?" The mayor turned to his son. "Is that true, Matt?"

"Me? I didn't! I wasn't..."

"DON'T LIE TO ME, BOY!" Now it was the mayor's turn to raise his voice.

Dana saw the look of desperation and bewilderment on Matt's face that only comes from a regular liar finding themselves not believed when they are telling the absolute truth. Matt looked at Professor Moreau, clearly knowing the professor was lying, but there was only completely convincing sincerity on the man's face. Both Matt's father and uncle were looking at Matt with disappointment in their eyes, and Matt scanned around frantically, looking for an ally somewhere... anywhere. Dana caught his eye, and without thinking, she stuck out her tongue.

"Come along, Dana. Let's go home." Her mother was steering Dana towards the car. She was suddenly very, very tired. She followed her mother and Noah, and as they reached the end of the manor house, she could hear the mayor yelling at his son, "How many times, *how* many times have I told you not to mess around with things you don't understand...!"

When they reached the parking lot, it so happened that Mrs. Dash had parked right next to Monera Lei, and the Lei family were also packing themselves into their car. Dana's mother put Chip's cage in the front seat, and opened the hatchback for Bosco, and Dana and Noah climbed into the back seat. From the back seat, Dana saw Lemma, also in the back seat of their car. Lemma rolled down her window and motioned to Dana to do the same. As her mom started the car, Dana rolled down her own window and raised her eyebrows in question.

““I can't believe your name will be on that trophy with mine,” Lemma said. “Your bacteria were all dying — your dumb project didn't even work. They gave you credit for an accident, made possible with tech you stole.”

Her sister Lambda leaned over and said, “That's what Technologists always do, use other people's work for their own gain.”

Dana paused a moment to collect her thoughts.

“What you're not getting about the STEM disciplines is the fundamental interconnectedness. Math is the root, Science the trunk, Engineering the branches, and Technology the fruit. The dependency — the causality — it goes both ways. Without Math, none of it exists. Without Technology, none of it has real purpose — it's all just academic games. Science doesn't happen without Math, but I notice you're not accusing scientists of using other people's math for their own gain.”

Noah leaned over from his side of the back seat. “If you wanted to know why I'm Team Tech and always will be, this is why. Technologists solve real-world problems. The moment you step out of the lab to use your knowledge in the real world, you've joined Team Tech.”

Lemma opened her mouth as if to respond, but no sound came out. She stared back at Dana and Noah, fish-faced.

“And if it bothers you to be on that trophy with me so much,” continued Dana, “think about this: the names will certainly be alphabetical, so my name will be first.”

Dana rolled up her window as her mom drove away, leaving Lemma sputtering in rage, while Bosco, looking back out the rear window, was panting in a manner that might have looked a lot like laughter.

As the car made its way out of the Cavor School driveway and onto Big Oak Drive, Noah and Dana each silently looked out their own window, lost in their own thoughts. Chip was wound into a sleepy red-orange ball in the front seat. Bosco climbed over the back seat and settled in between

Dana and Noah. He yawned a big, wide, doggy-yawn, and laid his head in Dana's lap. Dana's mother yawned long and loud, "Oh, excuse me," she said. "I was able to sleep a little on the plane, but airplane sleep doesn't count the same as real rest."

That goes double for spaceship sleep, thought Dana. She was having a hard time keeping her eyes open. Noah was already asleep in the seat next to her; She gazed up at the waning half-moon just rising in the night sky. It seemed to follow her, playing peekaboo behind the tree branches. She thought about what she had done up there, all the things she had experienced since first coasting into the Cavor School on her skateboard: aliens and robots and Lemma — *oh my*! She had even made more friends than enemies, and had worked together with her competitors to win. She looked out the window at the Hudson river, ebbing and flowing with the tide. How terrifying and fascinating and exciting it had all been, but also how good it felt to be back and safe and on her way to her own room and her own bed, where she could wake up like a regular kid for the first time in what felt like forever. She would be going to the Cavor School in the fall, and it was an even more remarkable place than she could have imagined. She had done it, and she could relax. About that part, anyway. There would be time enough to uncover all the mysteries at the Cavor School, and in the world, later, but for now — for now, she would let herself be lulled by the rhythm of the wheels, the quiet of the summer night, the slow, steady, sleeping breaths of her dog in her lap and her best friend beside her. Each time she blinked, her eyes stayed closed a little bit longer.

Finally, she said a silent goodnight to the Moon, and like a totally normal child, on a completely typical drive, on a summer night like so many others that was lovely in its very ordinariness, Dana fell asleep in the back of the car on the way home.

THE END

ACKNOWLEDGEMENTS

Thanks to the following people for taking the time to provide us with services not credited elsewhere, information, and/or feedback in an effort to help us make this book better, whether or not we were actually smart enough to use your work and/or listen to your recommendations:

Rascher M. Alcasid (and her North Bronx Writers Group) and Araceli Noriega (and her Bronx Writers Writing Group) helped this book achieve escape velocity. Janine Annett, Idria Barone Knecht, Andy Bomback, Calista Brill, Brent Britton, and Corinne Enright contributed essential expertise and advice. Kate Francia shared a lot about the craft of writing, and S. Lynne Fremont shared a lot about the craft of publishing. Phillip Gerba, Joan Gillman, Jo and Sean Hastings, Ash Kalb, John Lewis, Janet MacDonald, Alexander J. Marsh, Terry Marsh (K8NNU), and Perry Metzger (who denies responsibility for any scientific inaccuracies), contributed scientific, cultural, and technical expertise.

Dorothy Neagle, Stanley Rosenkampff, and Abigail Rudner pitched in with expertise and moral support throughout the process. Sonia Sobieski, Susan Tolman, Mitchell Tolman, and Sophie Williams (who will someday write books far better than this one) helped us translate a story about science into a human-readable book form.

ABOUT THE AUTHOR

S. S. Hudson is a pseudonym for person or persons, wishing to disguise their ethnicity, politics, race, religion, sex, species, and anything else interesting, so as not to distract from simple good intentions of making math, science, engineering, and technology interesting for girls, and boys, and small furry creatures from Alpha Centauri.

ABOUT ELEMENTIAD, INC.

Elementiad, Inc. is a privately held company based in New Hampshire and New York, USA. Elementiad creates games and stories grounded in math, science, technology, and engineering. Elementiad products emphasize building teams to make cool things and have fun. Founded in 2019 by a small group of technologists, writers, and an artist, Elementiad, Inc. is majority-owned and run by women and primary caregivers of school-age children.

MORE BOOKS BY S.S. HUDSON

The adventure continues in the next book:
Dana Dash — The Invisible Girl
https://www.danadash.com

MORE BOOKS BY ELEMENTIAD, INC.

Meet the Peach Mountain Friends
The Peach Orchard Project
Ten Snowy Days with Cat
Sid's Blossom Explosion
Monkey's Plans for Peaches
Leaf the Fun to Hannah

JOIN THE ELEMENTALIST ADVENTURE

https://www.elementiad.com